T0257200

5G Wireless Network Security and Privacy

5G Wireless Network Security and Privacy

Dongfeng (Phoenix) Fang
California Polytechnic State University, San Luis Obispo
San Luis Obispo

Yi Qian
University of Nebraska–Lincoln
Lincoln

Rose Qingyang Hu
Utah State University
Logan

Registered Offices

John Wiley & Sons, Inc., 111 River Street, Hoboken, NJ 07030, USA

John Wiley & Sons Ltd, The Atrium, Southern Gate, Chichester, West Sussex, PO19 8SQ, UK

For details of our global editorial offices, customer services, and more information about Wiley products visit us at www.wiley.com.

Library of Congress Cataloging-in-Publication Data applied for:

Hardback ISBN: 9781119784296

Cover Design: Wiley
Cover Image: © Immersion Imagery/Shutterstock

Set in 9.5/12.5pt STIXTwoText by Straive, Chennai, India
Printed and bound by CPI Group (UK) Ltd, Croydon, CR0 4YY

C9781119784296_021123

Contents

About the Authors

Dongfeng (Phoenix) Fang is an assistant professor in the Department of Computer Science and Software Engineering, and Department of Computer Engineering at California Polytechnic State University, San Luis Obispo (Cal Poly). Her research interests include network security, wireless security, security and privacy of Internet-of-Things, and security and privacy in machine learning.

Yi Qian, PhD, is an IEEE Fellow and is a Professor in the Department of Electrical and Computer Engineering at the University of Nebraska-Lincoln, USA.

Rose Qingyang Hu is a professor in the Department of Electrical and Computer Engineering and Associate Dean for Research of College of Engineering at Utah State University in Logan, USA. Her research interests include next-generation wireless communications, wireless network design and optimization.

Preface

5G wireless system is not only an evolution of the legacy 4G networks, but also a system with many new service capabilities, related to our daily life. To support these new service capabilities, 5G wireless systems integrate many new technologies, which can potentially bring new security vulnerabilities. Moreover, strict performance requirements for certain applications can not be satisfied with the current security solutions. For instance, vehicular communications over 5G require extremely low latency and IoT applications demand low overhead.

The new developments in network architecture and algorithms bring the challenges to the researchers to face new security vulnerabilities and high performance requirements of security solutions. This book intends to survey the current challenges in the field of security and privacy over 5G wireless systems, and to present flexible and efficient solutions for security and privacy over 5G wireless systems. Specifically, the book focuses on security and privacy improvements over 5G wireless systems based on security network architecture, cryptographic solutions, and physical layer security solutions for better flexibility and efficiency. There are seven chapters in this book.

Chapter 1 provides an introduction to 5G wireless systems. The chapter first introduces the motivations and objectives of 5G wireless networks. Based on the features of 5G wireless networks, 5G security drives and requirements are discussed. An overview of the 5G wireless network architecture is presented, and a comparison between the legacy cellular network and the 5G wireless network is discussed to better understand the systems.

Chapter 2 discusses cellular network security from 1G to 5G. A overview of security development from 1G to 4G is presented. Security attacks and security services in 5G wireless networks are discussed. Security architectures from 3G to 5G are illustrated.

Chapter 3 presents the security services and current solutions for security and privacy over 5G systems. The fundamental approaches for providing security in 5G wireless systems are first reviewed. Security solutions are introduced based on authentication, availability, data confidentiality, key management, and privacy.

Chapter 4 discusses interference management and security in heterogeneous networks (HetNet) over 5G wireless systems. Current studies and background of interference management and security issues on confidentiality are first introduced. A general HetNet system model and corresponding threat model are proposed. A security solution is proposed to utilize the existing interference to improve confidentiality in the 5G network.

This chapter presents the details of the proposed method. An experimental study and evaluation are then demonstrated.

Chapter 5 deals with improving flexibility and efficiency of security schemes for heterogeneous IoT networks over 5G systems. A few security and privacy schemes for IoT applications are first discussed. A general IoT system architecture, trust models, threat models, and design objectives are presented. An authentication and secure data transmission scheme is proposed. Security analysis is presented to verify the proposed scheme. This chapter also presents an experimental study and evaluation.

Chapter 6 explores the efficiency of secure mobility management over 5G networks based on software-defined networking (SDN). A HetNet system model is proposed over a SDN-based 5G network. The handover scenarios and procedures are discussed. The proposed authentication protocols are presented with security analysis and performance analysis and evaluations.

Chapter 7 discusses the open issues and possible future research directions over 5G wireless networks.

We hope that our readers will enjoy this book.

California
August 2022

Dongfeng (Phoenix) Fang
California Polytechnic State University, San Luis Obispo

Yi Qian
University of Nebraska-Lincoln

Rose Qingyang Hu
Utah State University

Acknowledgments

First, we would like to thank our families for their love and support.

We would like to thank our colleagues and students at California Polytechnic State University, San Luis Obispo, University of Nebraska-Lincoln, and Utah State University for their support and enthusiasm in this book project and topic.

We express our thanks to the staff at Wiley for their support and to the book reviewers for their great feedback. We would like to thank Sandra Grayson, Juliet Booker, and Becky Cowan for their patience in handling publication issues.

This book project was partially supported by the U.S. National Science Foundation under grants CNS-2007995, CNS-2008145, CCCS-2139508, and CCCS-2139520.

Introduction

The advanced features of fifth-generation (5G) wireless network systems yield new security and privacy requirements and challenges. This book addresses the motivation for security and privacy of 5G wireless network systems, an overview of 5G wireless network systems security and privacy in terms of security attacks and solutions, and a new security architecture for 5G systems. The aim of 5G wireless network security is to ensure the provision of robust security services to 5G wireless systems, without compromising the high-performance capabilities that characterize 5G technology. Due to the inadequacy of 4G security architectures for 5G systems, novel security architectures are required to ensure the effectiveness and adaptability of security in 5G wireless networks. The topics to be addressed in this book include:

- Introduction and background of 5G wireless networks,
- Attacks and security services in 5G wireless networks,
- Current 5G wireless security solutions,
- A new 5G wireless security architecture,
- Flexible and efficient security solutions, e.g., physical layer security, authentication, and mobility management.

1

Introduction to 5G Wireless Systems

Fifth-generation wireless networks, or 5G, are the fifth-generation mobile wireless telecommunications beyond the current 4G/International Mobile Telecommunications (IMT)-Advanced Systems [Panwar et al., 2016]. 5G wireless network is not only an evolution of the legacy 4G cellular networks but also a new communication system that can support many new service capabilities [Fang et al., 2017a]. In this chapter, we will introduce a general background of 5G wireless networks and 5G security, including motivations and objectives, security drives and requirements, and a general 5G wireless network architecture.

1.1 Motivations and Objectives of 5G Wireless Networks

The research and development of 5G technology is focused on achieving advanced features such as enhanced capacity to support a greater number of users at faster speeds than 4G, increased density of mobile broadband users to improve coverage [Xu et al., 2021], and supporting device-to-device (D2D) communications and massive machine-type communications [NGMN Alliance, 2015]. 5G planning also aims to provide better network performance at lower latency and lower energy consumption to better support the implementation of the Internet of Things (IoT) [Andrews et al., 2014]. More specifically, there are eight advanced features of 5G wireless systems as follows [Warren and Dewar, 2014]:

- Data rate: 1–10 Gbps connections to endpoints in the field;
- Low latency: 1-ms latency;
- Bandwidth: 1000× bandwidth per unit area;
- Connectivity: 10–100× number of connected devices;
- Availability: 99.999% availability;
- Coverage: 100% coverage;
- Network energy efficiency: 90% reduction of network energy usage;
- Device energy efficiency: Up to 10 years of battery life for low-power devices.

To achieve these eight advanced network performance features, various technologies [Agiwal et al., 2016] are applied to 5G systems, such as heterogeneous networks (HetNet), massive multiple-input multiple-output (MIMO), millimeter wave (mmWave) [Qiao et al., 2015], D2D communications [Wei et al., 2016], software-defined network (SDN) [Dabbagh

5G Wireless Network Security and Privacy, First Edition. Dongfeng (Phoenix) Fang, Yi Qian, and Rose Qingyang Hu.
© 2024 John Wiley & Sons Ltd. Published 2024 by John Wiley & Sons Ltd.

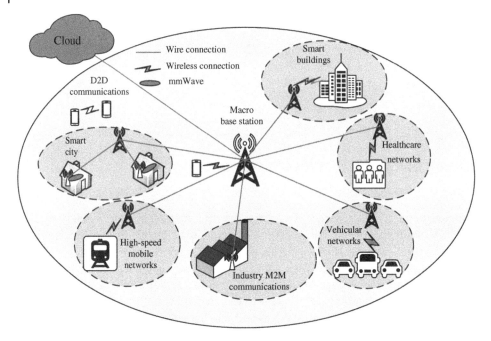

Figure 1.1 A generic architecture for 5G wireless systems.

et al., 2015], network functions virtualization (NFV) [Zhang et al., 2015], and networking slicing [NGMN Alliance, 2016]. The standardization process for 5G wireless systems has been carried out. Figure 1.1 illustrates a generic 5G wireless systems.

5G wireless systems can provide not only traditional voice and data communications but also many new use cases [Xu et al., 2022, Wang et al., 2021b], new industry applications, and a multitude of devices and applications to connect the society at large [AB Ericsson, 2018] as shown in Figure 1.1. Different 5G use cases are specified, such as vehicle-to-vehicle and vehicle-to-infrastructure communications [Fang et al., 2019b], industrial automation, health services, smart cities, and smart homes [Global Mobile Suppliers Association, 2015]. It is believed that 5G wireless systems can enhance mobile broadband with critical services and massive IoT applications [Qualcomm, 2016]. With the new architecture, technologies, and use cases in 5G wireless systems, it will face new challenges to provide security and privacy protections [Huawei, 2015].

1.2   Security Drives and Requirements

To accomplish the objectives of 5G wireless networks, several fundamental security drivers and requirements are necessary. Figure 1.2 illustrates the main drives for 5G wireless security as supreme built-in security, flexible security mechanisms, and automation. Supreme built-in security is needed since, in 5G, new use cases, new technologies, and new networking paradigms are introduced. The other use cases can introduce specific requirements, such as ultra-low latency in user communications, which will require improving the

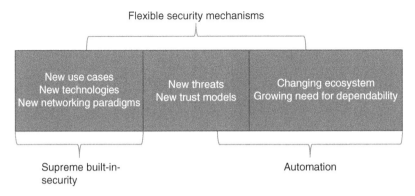

Figure 1.2 Security drives and requirements for 5G wireless security.

performance of the current security mechanisms. New technologies not only yield advanced service capabilities but also open the door to vulnerabilities and thus impose new security requirements in 5G [Liyanage et al., 2016]. In HetNet, different access technologies may have different security requirements, and a multi-network environment may need highly frequent authentications with stringent delay constraints [Wang et al., 2016b]. Massive MIMO has been deemed a critical 5G technique to achieve higher spectral efficiency and energy efficiency. It is also considered a valuable technique against passive eavesdropping [Deng et al., 2015]. Furthermore, SDN and NFV in 5G will support new service delivery models and thus require new security aspects [Chen et al., 2016b, Tian et al., 2017]. With the advent of 5G networking paradigms, a new security architecture is needed. To address these issues, security must be considered an integral part of the overall architecture and should initially be integrated into the system design.

To support various use cases, new technologies, new networking paradigms, new threats, new trust models in an optimal way, and flexible security mechanisms are needed with changing ecosystem and growing need for dependability. Based on the current research on 5G wireless networks, security services on 5G wireless networks have more specific requirements due to the advanced features that 5G wireless networks have, such as low latency, and high energy efficiency. With various applications on 5G wireless networks and their network performances, flexible security mechanisms are desired with better efficiency performance [Xu et al., 2019].

The trust models of the legacy cellular networks and 5G wireless networks are presented in Figure 1.3 [Huawei, 2015]. Not only full trust but also semi-trust or not trust are considered. Authentications are required not only between subscribers and the two operators (the home and serving networks) but also among service parties in 5G wireless networks. Moreover, for the use case of vertical industries, the security demands vary significantly among different applications. For instance, mobile devices require lightweight security mechanisms as their power resource constraint, while high-speed services require efficient security services with low latency. Therefore, the general flexibility for 5G security mechanisms is another critical requirement [Schneider and Horn, 2015]. Authentication management in 5G is more complex due to various types of and a massive number of devices connected. For different applications, different authentication models can be implemented. In Figure 1.3, user authentication can be done by the network provider, service provider, or both.

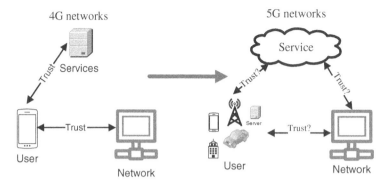

Figure 1.3 Trust model of 4G and 5G wireless networks.

Besides the supreme built-in security and flexibility security mechanisms, security automation is also a key element. It combines automated holistic security management with automated and intelligent security controls [NOKIA, 2017]. Since more personal information is used in various applications, such as surveillance applied over 5G wireless networks, privacy concerns escalate. Moreover, various services in 5G can be tied closer than before. For example, the fixed telephone line, internet access, and TV service can be terminated simultaneously due to the outage of a major network [Huawei, 2015]. Therefore, security automation is needed to make the 5G system robust against various security attacks.

1.3 5G Wireless Network Architecture

1.3.1 Overview of the 5G Wireless Network Architecture

The 5G wireless network architecture is introduced here. As shown in Figure 1.4, the illustrated general 5G wireless network architecture includes a user interface, a cloud-based heterogeneous radio access network, a next-generation core, distributed edge cloud, and a central cloud. The cloud-based heterogeneous radio access network can combine virtualization, centralization, and coordination techniques for efficient and flexible resource allocation. Based on different use cases, 3GPP classifies more than 70 different use cases into four different groups such as massive IoT, critical communications, network operation, and enhanced mobile broadband. In the cloud-based heterogeneous access network, besides the 3GPP access and non-3GPP access, other new radio technologies will be added for more efficient spectrum utilization. In the first stage of 5G, the legacy evolved packet core (EPC) will still be valid. Network slicing enables different parameter configurations for the next-generation core according to different use cases. New flexible service-oriented EPC based on network slicing, SDN, and NFV will be used in the next-generation core as virtual evolved packet core (VEPC) shown in Figure 1.4. The VEPC is composed of modularized network functions. Based on different use cases, the network functions applied to each VEPC can be various. In the VEPC, the control plane and user plane are separated for the flexibility and scalability of the next-generation core. Edge cloud is distributed to

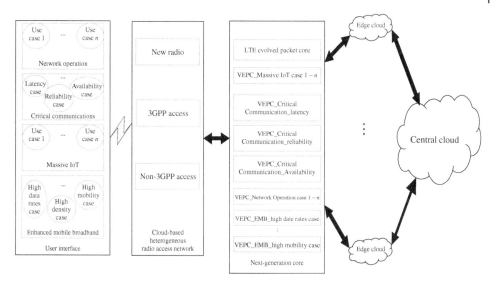

Figure 1.4 A general 5G wireless network architecture.

improve service quality. The central cloud can implement global data share and centralized control.

1.3.2 Comparison Between the Legacy Cellular Network and the 5G Wireless Network

Compared with legacy cellular networks, 5G wireless networks introduce some new perspectives and changes. (i) User equipment and services are not limited to regular mobile phones and regular voice and data services. Based on different use cases and requirements, user interfaces are classified into four different groups such as massive IoT, critical communications, network operation, and enhanced mobile broadband. Every use case can affect the radio access selection and VEPC functions. (ii) In addition to 3GPP access and non-3GPP access in the cloud-based heterogeneous radio access network, the 5G access network includes other new radios, which build the foundation of wireless standards for the next-generation mobile networks for higher spectrum utilization. The new radios can support the performance and connectivity requirements of various use cases in 5G wireless networks. Moreover, there are many technologies applied to the access network to improve the network performance, such as massive MIMO, HetNet, and D2D communications. (iii) The next-generation core will be based on the cloud using network slicing, SDN, and NFV to handle different use cases. The flexible service-oriented VEPC will be applied. With network slicing, SDN, and NFV, different network functions can be applied to the service-oriented VEPC for different use cases. The next-generation core is expected to be access-independent. Separation of control and user plane is important to achieve an access-agnostic, flexible, and scalable architecture. (iv) Edge cloud is applied to 5G wireless networks to improve the performance of the network, such as latency.

1.4 Conclusion

A general background of 5G wireless networks is introduced in this chapter. The motivations and objectives of 5G wireless networks are presented. With the expected improvements in 5G performance, security drives, and requirements are discussed. A general 5G wireless network architecture is illustrated in this chapter. Moreover, a comparison of a 5G wireless network architecture and legacy cellular network architecture is analyzed. It is clear that the 5G wireless network introduces significant flexibility to support new use cases and corresponding different service requirements. New security architecture and mechanisms are needed in 5G networks.

2

Security from Legacy Wireless Systems to 5G Networks

This chapter provides an introduction to the evolution of wireless network security, covering the security architecture and security services of the second generation (2G) to the fifth generation (5G) of wireless networks.

2.1 Network Security for Legacy Systems

Based on the technologies and network performances, vulnerabilities and security implementation from 2G to 4G are different.

Security services in 2G system such as global system for mobile communications (GSM) include user authentication, communication encryption, user anonymity, and detection of stolen/compromised equipment.

- User authentication: The user authentication is a challenge-response scheme between a user and the cellular network (such as visitor network and home network). To achieve user authentication requires a random number (RAND) and a key, which is pre-stored in the SIM card in the user mobile device. The SIM card also stores algorithms for achieving the user authentication. The user authentication is a one-way authentication, where the user is authenticated by the cellular network but the cellular network cannot be authenticated by the user. While this security service can effectively prevent users from misusing the network services, it cannot provide protection against rogue base stations.
- Communication encryption: After user authentication, a session key is generated to encrypt the user data between the user and the network on the radio link. A stream cipher is used in 2G GSM system.
- Anonymity: Anonymity is applied to provide privacy of international mobile subscriber identity (IMSI) of each user, since IMSI not and is also associated with user's identity. To achieve anonymity, instead of using IMSI all the time, a temporary mobile subscriber identity (TMSI) is used and updated between the user and the network based on different cases.
- Detection of stolen/compromised equipment: Each mobile device has an international mobile equipment identity, which can be used to achieve detection of stolen or compromised mobile device.

5G Wireless Network Security and Privacy, First Edition. Dongfeng (Phoenix) Fang, Yi Qian, and Rose Qingyang Hu.
© 2024 John Wiley & Sons Ltd. Published 2024 by John Wiley & Sons Ltd.

While 2G technology has established a foundation for security in cellular networks, it represents only the beginning of a continuous effort to enhance and strengthen the security measures in mobile communications. 2G networks are vulnerable to attacks targeting security algorithms, signaling networks (through exploitation of unencrypted messages), security protocols (such as rogue base station attacks), and denial-of-service attacks (including jamming). Furthermore, 2G does not provide data integrity.

3G such as universal mobile telecommunication system (UMTS) marks the beginning of a more comprehensive implementation of security measures in cellular networks. A security architecture is defined by 3GPP including five groups of security features in the UMTS. From a security perspective, 3G networks introduce significant improvements such as mutual authentication, two-way authentication, and key agreement protocols. In addition to these measures, 3G also offers enhanced data integrity compared to 2G. The introduction of stronger cryptographic algorithms further bolsters the security strength of 3G networks.

- AKA: The authentication and key agreement (AKA) mechanism involves three entities as a User Services Identity Module (USIM), the serving network, and the home network. A long term key is pre-shared between the USIM and the network. Based on a challenge–response mechanism, the network can authenticate the USIM, and the USIM can authenticate the home network. After the authentication, two keys will be generated to achieve data confidentiality and data integrity in the USIM and the network.
- Communication encryption: Confidentiality is provided in the 3G for data transmission over radio links between users and the base stations by encrypting the data with a cipher key, which is generated after authentication. A stream cipher is used with the cipher key, which is 128-bit long. There are also other inputs which will make sure that even for the same cipher key, the stream cipher can generate different keystream.
- Data integrity: Besides confidentiality, data integrity is provided in the 3G for data transmission over radio links between users and the base stations based on a message authentication code (MAC) with the integrity key of 128-bit long, which is generated after authentication.
- User identity confidentiality: As in 2G, preserving user identity confidentiality is a critical consideration in 3G networks. To achieve this, 3G networks implement temporary identities, such as TMSI in the circuit-switched domain and P-TMSI in the packet-switched domain. These temporary identities are used to limit the frequency of IMSI transmission and enhance user privacy.
- Detection of stolen/compromised equipment: Same as 2G.
- User-to-USIM authentication: A personal identification number (PIN) is used to achieve user-to-USIM authentication. This PIN is only known by the user and the USIM.

3G networks build upon the security mechanisms of 2G, while introducing modifications to enhance overall security. Although 3G expands network services and improves network performance, it also introduces new vulnerabilities, such as privacy concerns stemming from the introduction of location-based services. In summary, 3G represents a significant improvement in security compared to 2G.

4G long-term evolution (LTE) networks feature a different network architecture compared to 3G, designed to further improve network performance. This includes the use of evolved NodeBs (eNBs) among other enhancements. In 4G, control plane (CP) and user plane (UP) are separated. The security architecture of 4G networks differs slightly from that of 3G networks. In addition to further enhancing confidentiality and integrity algorithms, 4G networks introduce new security perspectives due to their different network architecture. For instance, 4G provides mutual authentication between user equipment (UE) and the evolved packet core (EPC), generating a cipher key and an integrity key after mutual authentication. However, in 4G, these keys are utilized differently. The use of comprehensive key management and handover security are also critical components of 4G security to ensure security in mobility scenarios. The following are the security services provided in 4G LTE system.

- 4G AKA: 4G AKA protocols build upon 3G AKA to provide mutual authentication between UE and EPC, ensuring the authentication of entities while also furnishing materials for UP radio resource control (RRC) and non-access stratum (NAS) cipher key and RRC and NAS integrity key.
- Key management: 4G has extended key hierarchy of 3G with more keys applied. The two keys generated after mutual authentication are used to derive an intermediate key K_{ASME} between UE and access security management entity (ASME). K_{ASME} can be used to generate more keys: $K_{NAS_{int}}$, $K_{NAS_{enc}}$, and K_{eNB}, where K_{eNB} can derive $K_{UP_{enc}}$, $K_{RRC_{int}}$, and $K_{RRC_{enc}}$.
- Confidentiality: Data confidentiality is provided in 4G by encrypting communications. $K_{NAS_{enc}}$ is used to provide data confidentiality between UE and mobility management entity (MME). $K_{UP_{enc}}$ and $K_{RRC_{enc}}$ are used to ensure communication confidentiality of user plane traffic and radio specific signaling, respectively.
- Integrity: Data integrity is ensured for communications between UE and MME by $K_{NAS_{int}}$. The integrity of radio specific signaling is protected by $K_{RRC_{int}}$. There is no data integrity protection for user plane traffic.
- Handover security: By introducing eNBs, 4G LTE also provides handover security between different eNBs to ensure the authentication and key management. Moreover, handover security between 4G and legacy systems is also provided. Many of the research work focused on improve the handover performance with required security.

4G networks boast stronger security implementations compared to the previous systems and continue to play a crucial role in modern cellular networks. Ongoing security enhancements are expected for 4G networks as well, further bolstering their security capabilities.

2.2 Security Attacks and Security Services in 5G Wireless Networks

2.2.1 Security Attacks

Security attacks can be categorized into two main types based on their attack features: passive attacks and active attacks [Stallings, 2017]. For a passive attack, attackers aim to learn or use information from legitimate users without influencing the communication itself. Two common types of passive attacks can occur in a cellular network: eavesdropping and traffic analysis. Passive attacks are intended to violate data confidentiality and

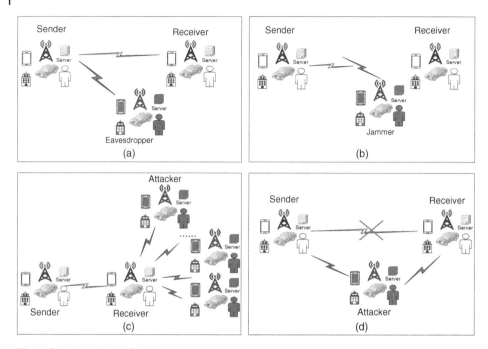

Figure 2.1 Attacks in 5G wireless networks (a) eavesdropping, (b) jamming, (c) DDoS, and (d) MITM.

user privacy. Unlike passive attacks, active attacks can include data modification or inter-ruption of legitimate communications. Typical active attacks are man-in-the-middle attacks (MITM), denial-of-service (DoS) attacks, and distributed denial-of-service (DDoS) attacks. Figure 2.1 illustrates both passive and active attacks in 5G wireless networks, each of which is introduced in three key aspects: the type of attack (passive or active), the security ser-vices implemented to mitigate the attack, and the corresponding methods utilized to avoid or prevent the attack.

- Eavesdropping: Eavesdropping is a passive attack that can be launched by an unintended receiver to intercept a message from others. Eavesdropping does not affect normal communication, as shown in Figure 2.1a. Due to its passive nature, eavesdropping is hard to be detected. Confidentiality is the security service applied to protect com-munication from this attack. Encryption of the signals over the radio link is the most common method used to mitigate the eavesdropping attack. The eavesdropper could not intercept the received signal directly since the data is encrypted. The encryption method heavily depends on the strength of the encryption algorithm, the length of the encryption key, and the computing capability of the eavesdropper. As computing power continues to rapidly advance and advanced data analysis technologies become increasingly prevalent, eavesdroppers can exploit these new capabilities to launch their attacks. The eavesdropping attack models in 5G are distinct from those in legacy systems, primarily due to the limited computing capabilities of the eavesdroppers in legacy system. Traditional mechanisms designed to combat eavesdropping are facing significant challenges, as many of them are predicated on the assumption of a small number of simultaneous eavesdroppers with low computing and data analysis capabilities. Furthermore, advanced technologies such as HetNet deployed in 5G wireless networks may exacerbate the difficulty of thwarting eavesdropping attacks.

- Traffic analysis: Traffic analysis is a type of passive attack where an unauthorized recipient intercepts information by analyzing the traffic of a received signal, without deciphering the content of the communication. The intercepted information can include details such as the location and identity of the communication parties, and can be used to deduce sensitive information about the communication. Even if the signal is encrypted, traffic analysis can still be utilized to reveal the patterns of communication parties. Traffic analysis attack does not impact legitimate communications either. Privacy protection is needed against traffic analysis. Currently, there is a significant amount of research focused on countering traffic analysis attacks, such as introducing artificial noise to disrupt traffic patterns and prevent the interception of sensitive information.
- Jamming: Unlike eavesdropping and traffic analysis, jamming can completely disrupt communication between legitimate users. Figure 2.1b is an example of the jamming attack. The malicious node can generate intentional interference that can disrupt the data communications between legitimate users. Jamming can also prevent authorized users from accessing radio resources. Availability is a security service aimed at mitigating the impact of jamming attacks.
- DoS and DDoS: DoS attacks, which aim to exhaust network resources, are a violation of network availability and can be launched by adversaries. Jamming can be employed as a means of carrying out a DoS attack. In the case where multiple distributed adversaries are involved, a DDoS attack can be formed, as shown in Figure 2.1c. These active attacks can occur at different network layers, and detection mechanisms are currently employed to identify and recognize DoS and DDoS attacks.
- MITM: In MITM attack, the attacker secretly takes control of the communication channel between two legitimate parties. The MITM attacker can intercept, modify, and replace the communication messages between the two legitimate parties. Figure 2.1d shows a MITM attack model. MITM is an active attack that can be launched in different layers. In particular, MITM attacks aim to compromise data confidentiality, integrity, and availability. Based on Verizon's data investigation report [Baker et al., 2011], MITM attack is one of the most common security attacks. In a legacy cellular network, a false base station-based MITM attack occurs when an attacker creates a fake base transceiver station to intercept and manipulate communications between a legitimate user and the intended destination, resulting in a compromised connection [Conti et al., 2016]. Mutual authentication between the mobile device and the base station is typically used to prevent false base station-based MITM.

2.2.2 Security Services

The emergence of 5G wireless networks, with their novel architecture, technologies, and use cases, introduces new security challenges and demands for security services. This subsection introduces four types of security services in 5G wireless systems: authentication (entity authentication, message authentication), confidentiality (data confidentiality, privacy), availability, and integrity.

2.2.2.1 Authentication

Entity authentication and message authentication are the two types of authentications. To tackle the previously discussed attacks, entity authentication and message authentication are essential in 5G wireless networks. Entity authentication is used to ensure the communicating entity is the one that it claims to be. In legacy cellular networks, mutual authentication between UE and MME is implemented before the two parties communicate. The mutual authentication between UE and MME is the most critical security feature in the traditional cellular security framework. The AKA in 4G LTE cellular networks is symmetric-key based. However, 5G requires authentication not only between UE and MME but also between other third parties, such as service providers. Given the distinct trust model of 5G networks compared to traditional cellular networks, a hybrid and adaptable approach to authentication management becomes necessary. Hybrid and flexible authentication of UE can be implemented in three different ways: authentication by the network only, authentication by the service provider only, and authentication by both network and service provider [Huawei, 2015]. Due to the very high-speed data rate and extremely low latency requirement in 5G wireless networks, authentication in 5G is expected to be much faster than ever. Moreover, the multi-tier architecture of the 5G may encounter very frequent handovers and authentications between different tiers in 5G. In Duan and Wang [2016], to overcome the difficulties of key management in HetNets and to reduce the unnecessary latency caused by frequent handovers, and authentications between different tiers, an software-defined networking (SDN)-enabled fast authentication scheme using weighted secure-context-information transfer is proposed to improve the efficiency of authentication during handovers and to meet 5G latency requirement. A public-key-based AKA mechanism is proposed in Eiza et al. [2016] and Zhang et al. [2017a] as a means of enhancing security services in 5G wireless networks.

Message authentication has become increasingly important with the various new applications in 5G wireless networks. Moreover, message authentication is facing new challenges with the more strict requirements on latency, spectrum efficiency (SE), and energy efficiency (EE) in 5G. In Dubrova et al. [2015] an efficient Cyclic Redundancy Check (CRC)-based message authentication for 5G is proposed as a means of detecting both random and malicious errors, without the need for increased bandwidth.

2.2.2.2 Confidentiality

Confidentiality encompasses two distinct dimensions, namely, data confidentiality and privacy. Data confidentiality protects data transmission from passive attacks by limiting the data access to intended users only and preventing access from or disclosure to unauthorized users. Privacy safeguards legitimate users against control or manipulation of their information. For example, privacy protects traffic flows from any analysis of an attacker. Traffic patterns in 5G networks can potentially reveal sensitive information, such as the location of senders or receivers. With a range of applications in 5G, including vehicle routing and health monitoring, user privacy can be compromised by vast amounts of data generated.

Data encryption is a widely adopted technique to ensure data confidentiality in which information is transmitted in such a way that unauthorized users cannot extract any meaningful information from it. Symmetric key encryption techniques can be applied to encrypt and decrypt data with one private key shared between the sender and the receiver. A secure key distribution method is required to share a key between the sender and the receiver. The conventional cryptography method is designed based on the assumption that attackers have limited computing capabilities. Rather than relying solely upon generic higher-layer cryptographic mechanisms, physical layer security (PLS) can support confidentiality service [Trappe, 2015] against jamming and eavesdropping attacks. Due to the large number of data involved in 5G networks, privacy protection services demand greater attention compared to legacy cellular networks [ERICSSON, 2015]. Anonymity service is an essential security requirement in many use cases. In many cases, privacy leakage can cause serious consequences. For example, health monitoring data reveals sensitive personal health information [Zhang et al., 2017a]; vehicle routing data can expose the location privacy [Eiza et al., 2016]. 5G wireless networks raise serious concerns about privacy leakage. In HetNets, due to the high density of small cells, the association algorithm can reveal users' location privacy. In Farhang et al. [2015], a differentially private algorithm is proposed to protect location privacy. In Abd-Elrahman et al. [2015], the proposed protocol secures privacy in group communications. In Eiza et al. [2016], cryptographic mechanisms and schemes are proposed to provide secure and privacy-aware real-time video reporting services in vehicular networks.

2.2.2.3 Availability

Availability refers to the extent to which a service is accessible and functional to legitimate users, regardless of their location or the time of request. Assessing the system's resilience against different types of attacks, including attacks to availability, is a critical performance metric in 5G networks. Availability, as a measure of system robustness, is particularly important in this context and is often targeted by active attackers. One of the major type of attacks on availability is the DoS attack, which can cause service access denial to legitimate users. Jamming or interference can disrupt the communication links between legitimate users by interfering with the radio signals. The robustness of 5G wireless networks is challenged by a multitude of unsecured IoT nodes, making them vulnerable to DDoS attacks that can significantly impact service availability.

Direct sequence spread spectrum (DSSS) and frequency hopping spread spectrum (FHSS) are two well-established PLS techniques employed to enhance availability. DSSS, originally developed for military applications in the 1940s, multiplies a pseudo-noise spreading code with the original data signal spectrum. Without knowledge of the spreading code, a jammer would require significantly more power to disrupt the legitimate transmission. For FHSS, a signal is transmitted by rapidly switching among many frequency channels using a pseudorandom sequence generated by a key shared between transmitter and receiver. Dynamic spectrum is applied to D2D communications and cognitive radio paradigm to improve the SE in 5G. In Adem et al. [2015], the authors observed that FHSS may lead to poor performance in the face of jamming attacks. A pseudorandom time hopping spread spectrum is proposed to improve jamming, switching, and error probability performance. Resource allocation is adopted to enhance the detection of the availability violation [Labib et al., 2015].

2.2.2.4 Integrity

Message authentication only verifies the message source and does not offer any safeguards against message duplication or modification. 5G networks aim to provide ubiquitous connectivity and support applications that are closely linked to daily human life, such as water quality monitoring and transportation scheduling. Data integrity is a critical security requirement for these specific applications.

Integrity ensures that the information is not tampered with or altered by active attacks launched by unauthorized entities. Malicious insider attacks, such as message injection or modification, can compromise data integrity. Insider attackers hold valid identities, making it challenging to detect their malicious activities. In use cases such as smart meters in smart grid [Yan et al., 2012], data integrity service needs to be provided against manipulation. Compared to voice communications, data can be more easily attacked and modified [SIMalliance, 2016]. Integrity services can be provided by mutual authentication, which can generate an integrity key. The integrity service of personal health information is required [Zhang et al., 2017a]. Authentication schemes can offer message integrity protection [Eiza et al., 2016].

2.3 The Evolution of Wireless Security Architectures from 3G to 5G

2.3.1 3G Security Architecture

The 3G security architecture defined by 3GPP TS 33.102 is shown in Figure 2.2. There are five groups of security features defined in the security architecture as follows:

- Network access security (I): This group offers a range of security features to ensure secure user access to 3G services and protect against radio access link attacks. These features encompass user identity confidentiality, user location confidentiality, user

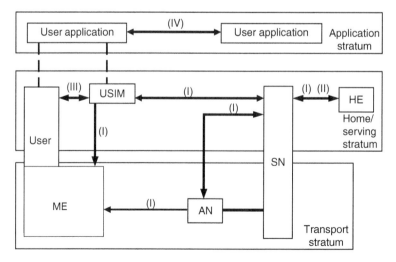

Figure 2.2 3G security architecture defined by 3GPP TS 33.102.

untraceability, user authentication, network authentication, data confidentiality, data integrity, and mobility equipment identification.

- Network domain security (II): The security feature set offered in this group aims to facilitate secure exchange of signaling data among users and safeguard against attacks on the wired network part of the 3G wireless system.
- User domain security (III): This group defines a set of security features to ensure secure mobile station access, including user-to-USIM authentication and USIM-terminal authentication.
- Application domain security (IV): The set of security features defined in this group focuses on facilitating secure communication between applications in both user and provider domains.
- Visibility and reconfigurability of security (V): This group defined a set of features to inform users about the availability of security features and whether the provision and use of services depend on these security features.

2.3.2 4G Security Architecture

The 4G network employs both 3GPP access and non-3GPP access to connect the evolved packet system (EPS). The security architecture of 4G is similar to that of 3G, with five security groups defined, as shown in Figure 2.3.

- Network access security (I): The set of security features is designed to ensure user secure access services provided by the 3GPP EPC. To achieve this goal, various security services, including authentication, key management, data confidentiality, data integrity, and user identity confidentiality, are provided.
- Network domain security (II): Similar to 3G, this security feature set aims to ensure secure exchange of signaling data between users and protection against attacks on the wired network part of the 4G wireless system.

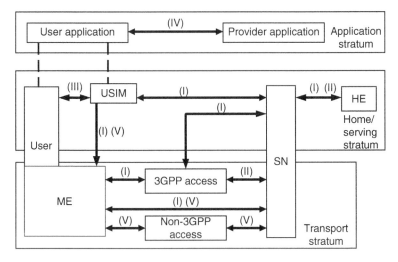

Figure 2.3 4G security architecture defined by 3GPP TS 33.402.

- User domain security (III): Same as the user domain security in 3G, in 4G, the set of security features in this group ensures secure access to the mobile station.
- Application domain security (IV): Similar to 3G, the set of security features in this group aims to ensure secure messaging between applications in both the user and provider domains.
- Non-3GPP domain security (V): This group focuses on providing security features specifically for non-3GPP access.

2.3.3   5G Wireless Security Architecture

2.3.3.1   Overview of the Proposed 5G Wireless Security Architecture

In Fang and Qian [2020], the authors proposed a new 5G wireless security architecture as illustrated in Figure 2.4. With the new characteristics of the next-generation core, a separation of the data plane and control plane of virtual evolved packet core (VEPC) is proposed, where the data plane can be programmable for its flexibility. The major network functions in the control plane of the next-generation core are identified in TR 23.799, which are utilized in this security architecture as follows:

- Access and mobility management function (AMF): The function serves to manage control access and mobility, and it is implemented in the MME of legacy cellular networks. However, its application may vary depending on the specific use case. Notably, the mobility management function is not required for fixed-access applications.
- Session management function (SMF): This function is capable of establishing and managing sessions according to network policies. Additionally, in cases where a single user has multiple sessions, different SMFs can be assigned to manage each session while still being associated with a single AMF.
- Unified data management (UDM): UDM manages subscriber data and profiles (such as authentication data of users) for both fixed and mobile access in the next-generation core.
- Policy control function (PCF): This function provides roaming and mobility management, quality of service, and network slicing. AMF and SMF are controlled by PCF. Differentiated security can be provided with PCF.

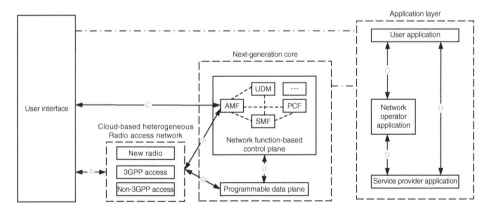

Figure 2.4   The 5G wireless security architecture.

AMF and SMF have integrated into the legacy cellular networks as MME. The separation of AMF and SMF can support a more flexible and scalable architecture. In the network function-based control plane, different network functions can be applied to different use cases.

2.3.3.2 Security Domains

Similar to the legacy cellular networks, four security domains are defined in Figure 2.4 as A, B, C, D. The details of these security domains are introduced as follows.

Network access security (A): The set of security features that enable secure user interface access to the next-generation core and safeguard against potential attacks on the radio access link. The new physical layer technologies applied to the radio access network, including massive multiple-input and multiple-output (MIMO), HetNet, D2D communications, and mmWave, bring new challenges and opportunities in network access security. The security of this level is reinforced by various mechanisms, including confidentiality and integrity protection that are implemented between the user interface and radio access network. Ongoing research on network access security is mainly focused on ensuring the confidentiality of user identity and location, as well as the confidentiality of user and signaling data. In addition, entity authentication is also an area of interest in this field.

Network domain security (B): The set of security features that safeguard against the attacks on the wireline network part of the 5G system and facilitate secure exchange of signaling and user data among entities and functions. As shown in Figure 2.4, this level of security is present between an access network and next-generation core, control plane, and user plane. Since new technologies such as cloud computing, network slicing, and network functions virtualization (NFV) are applied to 5G core and radio access networks, new vulnerabilities at this level need to be addressed. However, with the separation of the control plane and user plane, the amount of signaling data will be significantly reduced. The network function-based control plane also reduces the required signaling overhead for data synchronization. Entity authentication, data confidentiality, and data integrity are the main security services at this level. With the independent characteristics of access techniques of AMF, the network domain security performance can be simplified and improved.

User domain security (C): The set of security features that establish mutual authentication between the user interface and the next-generation core, ensuring secure access to the control plane from the user interface. Authentication is the main focus at this level. Based on the use case, authentication may be needed for more than two parties. For example, authentication can be required between the user and a network operator as well as between the user and a service provider. Moreover, different service providers may need to authenticate each other to share the same user identity management. Compared to device-based identity management in legacy cellular networks, new identity management methods are needed to improve security performance.

Application domain security (D): The set of security features that guarantee secure message exchange between applications at the interfaces, between the user interface and service provider, and between the user and network operator.

2.4 Summary

This chapter provides an introduction to wireless network security from legacy cellular systems to 5G systems, followed by a discussion on security attacks and expected security services in 5G wireless systems. Finally, the chapter gives an overview on the evolution of wireless security architectures from 3G to 5G systems.

3

Security Mechanisms in 5G Wireless Systems

This chapter discusses the security mechanisms in 5G wireless systems, focusing on providing different security services such as authentication, availability, data confidentiality, key management, and privacy. These mechanisms can be categorized into two main approaches: cryptographic with new networking protocols and physical layer security (PLS). Before delving into the 5G security solutions, the basics of these two categories are introduced.

3.1 Cryptographic Approaches and Physical Layer Security

Cryptographic techniques are widely used security mechanisms in 5G wireless networks and are typically implemented at the upper layers with new networking protocols. Modern cryptography consists of symmetric-key cryptography and public-key cryptography. Symmetric-key cryptography refers to the encryption methods in which a secret key is shared between a sender and a receiver. Public-key cryptography, also known as asymmetric cryptography, involves the use of two different keys: a public key for encryption and a private key for decryption. Besides cryptographic approaches, PLS is also widely explored in 5G systems.

This section outlines security solutions based on cryptographic approaches, including symmetric-key cryptography and public-key cryptography. Table 3.1 provides an overview of the various security services discussed for cryptographic approaches.

- Symmetric-key cryptography: Symmetric-key cryptography refers to any cryptographic algorithm that uses a shared secret key to encrypt and decrypt data. It can provide data confidentiality, authentication, and data integrity [Qian et al., 2021]. Figure 3.1 illustrates how symmetric-key-based encryption and decryption can be used to achieve confidentiality by using the same key for both processes. This approach is commonly employed to ensure data confidentiality for data communication and storage. Encryption and decryption use the same key. Symmetric-key-based encryption and decryption are widely used to ensure data confidentiality for data communication and storage.

  In addition to symmetric-key-based encryption and decryption, message authentication code (MAC) and hash functions can also be utilized to achieve message authentication and integrity.

5G Wireless Network Security and Privacy, First Edition. Dongfeng (Phoenix) Fang, Yi Qian, and Rose Qingyang Hu.
© 2024 John Wiley & Sons Ltd. Published 2024 by John Wiley & Sons Ltd.

Table 3.1 Cryptography and physical layer security.

Security mechanisms	Related attacks	Security services
Symmetric-key cryptography	Eavesdropping, message modification, reply	Data confidentiality, data integrity, authentication.
Public-key cryptography	Eavesdropping, message modification, reply, masquerade	Data confidentiality, data integrity authentication, nonrepudiation.
Physical layer security	Eavesdropping, traffic analysis, jamming	Data confidentiality, authentication, availability.

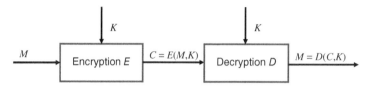

Figure 3.1 Symmetric-key encryption and decryption.

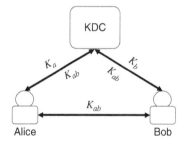

Figure 3.2 Key management with KDC.

Symmetric-key cryptography is based on the assumption that the sender and the receiver both securely share and store a secret key K. Generally, K can be distributed by a key distribution center (KDC), which is fully trusted by both the sender and receiver. The KDC-based system suffers a single point-of-failure issue. The key management based on KDC is shown in Figure 3.2. KDC has a pre-shared key with each user associated with it, such as Alice shares K_a and Bob shares K_b with KDC. Based on K_a and K_b, KDC can generate a key K_{ab} and distribute this key securely to Alice and Bob to secure the communications between Alice and Bob for each session.

- Public-key cryptography: Public-key cryptography utilizes a pair of keys – a public key and a private key – to achieve confidentiality, authentication, integrity, and non-repudiation. This cryptographic algorithm eliminates the need to securely share the key between parties [Stallings et al., 2012].

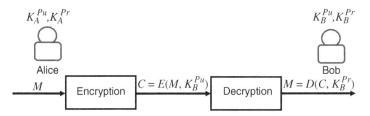

Figure 3.3 Public-key-based encryption and decryption.

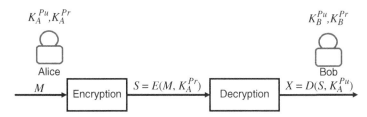

Figure 3.4 Public-key-based digital signature.

Figure 3.3 shows public-key-based encryption and decryption. Alice has her public key K_A^{Pu} and private key K_A^{Pr}, and Bob has his public key K_B^{Pu} and private key K_B^{Pr}. Alice and Bob possess each other's public keys. Alice can use Bob's public key to encrypt the plaintext M to the ciphertext C, and Bob can use his private key to decrypt the ciphertext C to the plaintext M. This method can also be used for key exchange and distribution.

Besides confidentiality and key exchange, public-key cryptography can also provide a digital signature, which ensures data integrity, message authentication, and non-repudiation. Different from confidentiality, digital signature scheme is illustrated in Figure 3.4. Alice can sign a message using her private key to generate a digital signature. Bob can verify the digital signature by decrypting the signed message using Alice's public key.

Public-key cryptography is commonly utilized for secret key distribution and providing security services like non-repudiation, which cannot be accomplished by symmetric-key cryptography. Compared to symmetric-key cryptography, key management is relatively straightforward in public-key cryptography, where users share their public keys, and a certificate authority signs all users' public keys.

The objective of PLS is to leverage the physical properties of communication channels to establish secure communication. To improve communication security at the physical layer in wireless networks, several methods have been developed, which can be classified into five major categories: theoretical security capacity, channel, coding, power, and other approaches [Shiu et al., 2011]. PLS techniques primarily focus on preventing eavesdropping and jamming attacks, as illustrated in Table 3.1.

Theoretical security capacity research primarily centers around the analysis of secrecy capacity [Oggier and Hassibi, 2011], which is the disparity in data rates between an authorized user and an unauthorized user. The secrecy capacity C_s is defined as:

$$C_s = C_m - C_e, \tag{3.1}$$

where C_m is the main channel capacity of the legitimate user, and C_e is the channel capacity of the eavesdropper. Besides secrecy capacity, secrecy outage probability is defined as the instantaneous secrecy capacity is less than a target secrecy rate R_t, where $R_t > 0$, and:

$$P_{out}(R_s) = P(C_s < R_t). \tag{3.2}$$

PLS primarily consists of three distinct approaches:

- Channel approaches leverage the unique characteristics of the communication channel to enhance security. These methods can be broadly categorized into three types: radio frequency (RF) fingerprinting [Sperandio and Flikkema, 2002], algebraic channel decomposition multiplexing (ACDM) precoding [Li and Ratazzi, 2005], and randomization of multiple-input and multiple-output (MIMO) transmission coefficients [Abbasi-Moghadam et al., 2008].
- Coding approaches: Coding approaches are typically employed to enhance resilience against eavesdropping and jamming attacks. Error-correcting coding and spread spectrum coding are two commonly used methods in coding approaches.
- Power approaches: Power approaches can also enhance data protection. The advancements in hardware and software in 5G have made it possible to employ techniques such as directional antennas, artificial noise injection, and power control to mitigate eavesdropping and jamming attacks. Directional antennas allow legitimate users to receive data from a specific direction, while eavesdroppers are unable to receive the signal. Artificial noise injection can also enable secure communication by decreasing the signal-to-noise ratio.

3.2 Authentication

Authentication is a critical security service in 5G wireless networks. In previous cellular networks, authentication schemes were typically based on symmetric-key cryptography. Implementing authentication schemes can provide several security benefits. In 3G cellular networks, mutual authentication is implemented between a mobile station and the network. Following the authentication, a cipher key and an integrity key are generated to provide both data confidentiality and data integrity between the mobile station and the base station (BS).

Authentication in 5G systems poses various challenges such as meeting low latency requirements, supporting device-to-device (D2D) communication, ensuring efficient utilization of bandwidth and energy, addressing limited computational capabilities of devices, and considering new security service requirements.

- Low latency requirement: Due to the low latency requirements of 5G networks, authentication schemes must be more efficient than ever before. To leverage the advantages of software-defined networking (SDN), in Duan and Wang [2016], a fast authentication scheme in SDN is proposed, which uses weighed secure-context-information (SCI) transfer as a non-cryptographic security technique to improve authentication efficiency during frequent handovers in a HetNet in order to address the latency requirement. The proposed fast authentication protocol includes full authentication and weighted SCI transfer-based fast authentication. As shown in Figure 3.5, after the first full authentication in one cell, it can be readily applied in other cells with MAC address verification, which only needs local processing. Moreover, full authentication can even be done without disrupting user communication. A valid time duration parameter is used to flexibly adjust the secure level requirement. Compared to cryptographic authentication methods, the proposed method is less susceptible to be compromised as it relies on the user's inherent physical layer attributes. SCI utilizes multiple physical layer characteristics to enhance authentication reliability for applications that require a high level of security. The SDN-enabled authentication model is illustrated in Figure 3.5. The SDN controller employs a model to monitor and predict the user's location, allowing it to prepare the relevant cells before the user's arrival and achieve seamless handover authentication. Physical layer attributes are used to provide unique fingerprints of the user and to simplify the authentication procedure. Three types of physical layer attributes are utilized as fingerprints for individual users. Once authentication is complete, the validated initial attributes are acquired. The SDN controller constantly gathers observations by sampling various physical layer attributes from received packets to validate the original attributes. The original file and observation results both contain the mean and variance of the selected attributes. Then the mean attribute offset can be calculated based on the validated original attributes and observed attributes. If the attribute offset is less than a pre-determined threshold, the user equipment is considered legitimate. To evaluate the performance of the authentication method, an SDN network model using priority queuing is proposed. The arriving traffic is being modeled using a Pareto distribution. The delay in authentication is being compared across multiple network utilization scenarios. The simulation results compared the delay performance of SDN-enabled fast authentication with that of the conventional cryptography authentication method. SDN-enabled fast authentication exhibits superior delay performance, leveraging the flexibility and programmability offered by SDN in 5G networks.

- Supporting D2D communication: To mitigate the security concerns stemming from the absence of a security infrastructure for D2D communications, [Abualhaol and Muegge, 2016] introduced a security scoring mechanism based on continuous authenticity. This approach employs the concept of legitimacy patterns to implement continuous authenticity, allowing for the detection of attacks and the measurement of system security scores. The legitimacy pattern involves the insertion of a redundant sequence of bits into a packet to facilitate attack detection. The simulation results demonstrate the efficacy of the proposed security scoring using legitimacy patterns. Furthermore, the authors suggest that taking into account both technical aspects and human behavior could enhance the performance of legitimacy patterns.

- High efficiency in bandwidth utilization and energy consumption: In their work [Dubrova et al., 2015], the authors introduced a new message authentication mechanism based on cyclic redundancy check (CRC) that combines the high security and exceptional efficiency in bandwidth utilization and energy consumption achievable in 5G networks. Notably, this scheme is capable of detecting double-bit errors within a single message, enhancing the reliability and accuracy of the authentication process. The CRC codes-based cryptographic hash functions are defined. A linear feedback shift register (LFSR) is used to efficiently implement the CRC encoding and decoding. The message

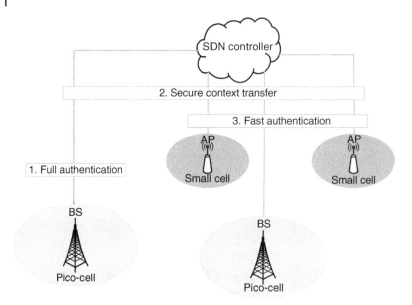

Figure 3.5 A SDN-enabled authentication model. Source: Adapted from Duan and Wang [2016].

authentication algorithm produces an authentication tag based on a secret key and the message. The security assumption underlying this scheme is that the adversary possesses the family of hash functions but lacks access to the specific polynomial $g(x)$ and pad s utilized for generating the authentication tag. The generator polynomial is changed periodically at the beginning of each session, and pad s is changed for every message. The new family of cryptographic hash functions based on CRC codes with generator polynomials in $g(x) = (1 + x)p(x)$ are introduced, where $p(x)$ is a primitive polynomial. While maintaining the implementation simplicity of cryptographically non-secure CRCs, the proposed CRC incorporates a re-programmable LFSR that necessitates adaptable connections.

- Limited computational capability devices: Radio frequency identification (RFID) has been widely adopted across numerous industries and applications. Due to various limitations in low-cost RFID tags, the encryption algorithms and authentication mechanisms applied to RFID systems need to be very efficient. Simple and fast hash function is considered for the authentication mechanisms of RFID systems. When using a single RFID tag for multiple applications, it is important to take into account the possibility of revocation in the authentication scheme. In Fan et al. [2015], the authors proposed a revocation method in the RFID secure authentication scheme in 5G use cases. A hash function and a random number are used to generate the corresponding module through a typical challenge–response mechanism. Figure 3.6 shows an authentication process of the RFID secure application revocation scheme. The reader contains a pseudo-random number generator (PNG), and the server holds a hash function and a database (HFD). The server establishes a tag record for each legitimate tag as (IDS, ID_i) and a group of corresponding application records as $(K_{i,j}^{old}, K_{i,j}^{now})$. q is the authentication request generated by the reader. r_1 is the first random number generated by the PNG in the

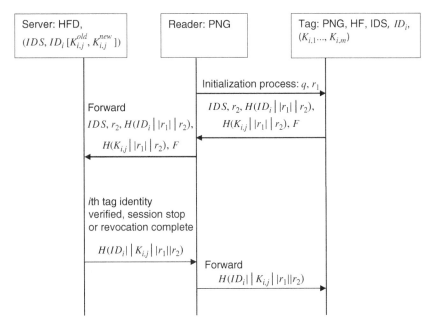

Figure 3.6 The authentication process of the RFID secure application revocation scheme. Fan et al. [2015].

reader. Upon receiving the authentication request, the tag generates the second random number r_2 and calculates two hash authentication messages M_1, M_2, and the exclusive or (XOR) authentication information value $F = E \oplus K_{i,j}$, where E is the current value of the status flag information, which is used to determine whether to revoke or to certify the application. The presented security and complexity results demonstrate that the proposed scheme offers a better security and the same level of complexity as existing schemes.

- New security services consideration: Considering the open nature of D2D communications between medical sensors and the high privacy requirements of the medical data, in

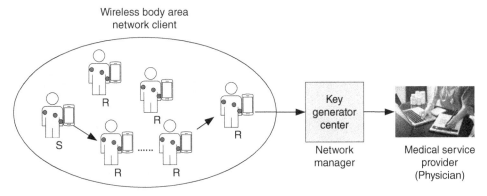

Figure 3.7 A m-health system model Zhang et al. [2017a]. Source: Adapted from NINENII/Adobe Stock Photos.

Zhang et al. [2017a], by utilizing certificate-less generalized signcryption (CLGSC) technique, the authors proposed a light-weight and robust security-aware (LRSA) D2D-assist data transmission protocol in an m-health system. The m-health system is modeled in Figure 3.7, where S is the source node and R represents the relay node. Anonymous and mutual authentication are implemented between the client and the physician in a wireless body area network to protect the privacy of both the data source and the intended destination. The signcryption of the message μ_S and encryption of its identity e_H^S are applied to the source client to authenticate the physician. A certificated-less signature algorithm is applied to the source client data before it is sent out. The source data identity can only be recovered by the intended physician who has the private key (x_H, z_H). The ciphertext μ_S should be decrypted after the source identity is recovered with the right session key. Therefore, even if the private key is leaked out, without the session key, the ciphertext is still safe. On the other hand, by verifying the signcryption μ_S, the physician can authenticate the source client. The relay nodes can verify the signature and then forward the data with their own signatures. The computational and communication overheads of the proposed CLGSC are compared with the other four schemes. Simulation results show that the proposed CLGSC scheme has a lower computational overhead than the other four schemes.

Compared to IEEE 802.11p and legacy cellular networks, 5G is a promising solution to provide real-time services for vehicular networks. However, security and privacy need to be enhanced in order to ensure the safety of transportation system. In Eiza et al. [2016], a reliable, secure, and privacy-aware 5G vehicular network supporting real-time video services is presented. The system architecture is shown in Figure 3.8, which includes a mobile core network (MCN), a trusted authority (TA), a department of motor vehicles

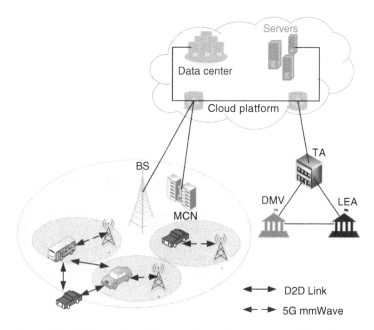

Figure 3.8 A 5G-enabled vehicular network. Source: Adapted from Eiza et al. [2016].

(DMV), and a law enforcement agency (LEA). D2D communications and mmWave techniques are adopted in 5G vehicular communications. As shown in Figure 3.8, HetNet is applied to expand network capacity and achieve high user data rates. The cloud platform provides massive storage and ubiquitous data access. The proposed cryptographic mechanisms include a pseudonymous authentication scheme, public-key encryption with keyword search, ciphertext-policy attribute-based encryption, and threshold schemes based on secret sharing. The pseudonymous authentication scheme with strong privacy preservation [Sun et al., 2010] is applied to optimize the certification revocation list size, which is in a linear form with respect to the number of revoked vehicles so that certification verification overhead is the lowest. The authentication requirements include vehicle authentication and message integrity, where vehicle authentication allows the LEA and official vehicles to check the sender's authenticity. Authentication is achieved by using a public-key-based digital signature that binds an encrypted traffic accident video to a pseudonym and to the real identity of the sender. The pseudonymous authentication technique can achieve the conditional anonymity and privacy of the sender.

3.3   Availability

Availability is a key metric to ensure ultra-reliable communications in 5G. By emitting wireless noise signals randomly, a jammer can significantly degrade the performance of mobile users and even completely block the availability of services. Jamming is one of the typical mechanisms used to launch DoS attacks in the physical layer. Most of the anti-jamming schemes use the frequency-hopping techniques, in which users hop over multiple channels to avoid the jamming attack and to ensure the availability of services. The main challenge with the advent of new technologies and use cases in 5G is to improve the performance of service availability.

In Li et al. [2011], the authors proposed a secret adaptive frequency-hopping scheme as a possible technique against DoS based on a software-defined radio platform in 5G system. The proposed bit error rate (BER) estimator based on physical layer information is applied to decide frequency blacklisting under a DoS attack. Dynamic spectrum access users may not be able to efficiently use the frequency-hopping technique due to the high switching rate and probability of jamming, as it requires access to multiple channels.

To reduce the switching rate and probability of jamming, in Adem et al. [2015], a pseudorandom time-hopping anti-jamming scheme is proposed for cognitive users in 5G to countermeasure jamming attacks. A cognitive jammer with limited resources is included in the model to analyze the impact of spectrum dynamics on the performance of mobile cognitive users. Analytical solutions of jamming probability, switching rate, and error probability are presented in the paper. The jamming probability, which is related to delay performance and error probability, is lower when the jammer has limited access opportunities. Compared to the frequency-hopping system, the time-hopping system exhibits superior switching probability. When using the same average symbol energy per joule, time-hopping has a lower error probability than frequency-hopping. However, the performance gain reaches a saturation point at a certain symbol energy level. The authors

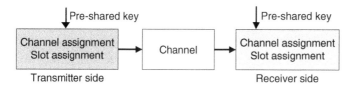

Figure 3.9 A pseudorandom time-hopping system block diagram. Source: Adapted from Adem et al. [2015].

highlighted the suitability of the proposed time-hopping technique for D2D links in 5G wireless networks due to its strong performance in terms of energy efficiency and spectral efficiency. Additionally, this technique has the ability to provide jamming resilience with minimal communication overhead. However, a pre-shared key is required for the time-hopping anti-jamming technique. Figure 3.9 illustrates the block diagram of the pseudorandom time-hopping system. Similar to frequency hopping, time hopping also necessitates a pre-shared key to determine the hopping sequence.

In order to protect nodes with limited computational capabilities from malicious radio jamming attacks in a 5G wireless network, a fusion center was employed in Labib et al. [2015]. As an exercise in strategic resource distribution, a non-cooperative Colonel Blotto game is formulated between the jammer and the fusion center. Figure 3.10 shows the resource allocation model between the fusion center and the malicious jammer. The jammer aims to jeopardize the network without getting detected by distributing its power among the nodes intelligently. On the other hand, the fusion center as a defender aims to detect such an attack by a decentralized detection scheme at a certain set of nodes. The fusion center can allocate more bits to these nodes for reporting the measured interference. Once the attack is detected, the fusion center will instruct the target node to increase its transmit power to maintain a proper signal to interference

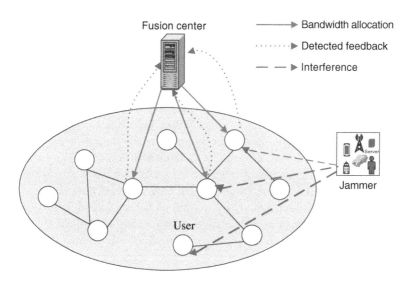

Figure 3.10 The resource allocation model. Source: Adapted from Labib et al. [2015].

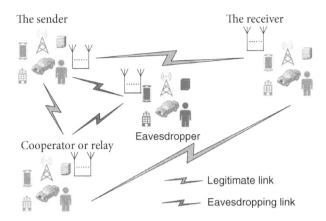

The sender

The receiver

Eavesdropper

Cooperator or relay

Legitimate link

Eavesdropping link

Figure 3.11 A general system model with eavesdropping attacks.

and noise ratio (SINR) for normal communications. Simulation results demonstrate a notable improvement in error rate performance as the fusion center has more bits to allocate among nodes. Moreover, the proposed resource allocation mechanism exhibits superior performance compared to the mechanism that randomly assigns the available bits.

3.4 Data Confidentiality

To mitigate eavesdropping attacks, data confidentiality services are typically necessary. The general system model with eavesdropping attacks is shown in Figure 3.11. In this section, we discuss data confidentiality based on power control, relay, artificial noise, signal processing, and cryptographic methods.

- Power control: The objective of power control for security is to regulate the transmit power in such a way that the signal cannot be recovered by the eavesdropper. Based on the most simple eavesdropping attack model with a single eavesdropper armed with a single antenna, in Ghanem and Ara [2015], the authors proposed a distributed algorithm to secure D2D communications in 5G system, which allows two legitimate senders to select whether to cooperate or not and to adapt their optimal power allocation based on the selected cooperation framework. Figure 3.11 shows a general system model with eavesdropping attacks. In the system model in Ghanem and Ara [2015], the sender, relay or cooperator, receiver, and eavesdropper are named as Alice, John, Bob, and Eve, respectively. Each user has a single antenna. A bidirectional link connects Alice and John. The problem is formulated to maximize the achievable secrecy rates for both Alice and John as follows [Ghanem and Ara, 2015]

$$C_a = \max\left(R_{ajb} - R_{ae}\right),\tag{3.3}$$

$$\text{s.t.}\, P_j + P_{jb} \leq P_J,\tag{3.4}$$

$$C_j = \max (R_{jab} - R_{je}), \tag{3.5}$$

$$\text{s.t. } P_a + P_{ab} \le P_A, \tag{3.6}$$

where C_a and C_j represent the secrecy rates of Alice and John, respectively. R_{ajb} and R_{jab} are the achievable rates of Alice and John, respectively, helping to relay data for each other. R_{ae} and R_{je} are the achievable rates of eavesdroppers from Alice and from John, respectively. Equations (3.4) and (3.6) represent the transmit power limitation of the two legitimate senders. Two cooperation scenarios are considered, namely cooperation with a relay and cooperation without a relay. In cooperation with the relay scenario, Alice and John can help relay data to each other using the shared bi-directional link. In cooperation without relay, Alice and John coordinate their respective transmission power to maximize the secrecy rate of the other one. The optimization problem in the noncooperative scenario is presented as a point of comparison. The distance between the legitimate transmitter and the eavesdropper is constrained to avoid distance attacks as the eavesdropper may have a better quality of received signal on the transmitted message than the legitimate receiver. Simulation results show that the achievable secrecy rates of Alice and John are improved by relaying data for each other. With the increase in distance between the transmitter and the receiver, the benefit from cooperation decreases, and at some point, non-cooperation could become more beneficial to the legitimate transmitter.

With no relay or cooperation, based only on power control and channel access, in Luo et al. [2015b], the authors developed a Stackelberg game framework for analyzing the achieved rate of cellular users and the secrecy rate of D2D users in 5G by using PLS. The system model includes one BS, a number of cellular users, one D2D link, and one eavesdropper, as shown in Figure 3.12. The utility function of cellular user achieved rates and D2D user secrecy rates are expressed as functions of channel information and transmission power [Luo et al., 2015b]:

$$u_{c,i} = \log_2(1 + SINR_{c,i}) + \alpha\beta P_D h_{dc}, \tag{3.7}$$

$$u_d = [\log_2(1 + SINR_d) - \log_2(1 + SINR_e)] - \alpha P_D h_{dc}, \tag{3.8}$$

where α is the price factor and β is the scale factor. The first term in $u_{c,i}$ represents the data rate of the ith cellular user, and the second term compensates for the interference from the D2D link, where P_D is the transmit power of the D2D user and h_{dc} is the channel gain from

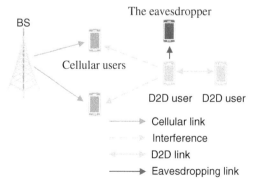

Figure 3.12 The system model with D2D link and an eavesdropper. Source: Adapted with Luo et al. [2015b].

the D2D user to cellular users. The utility function of D2D users includes the secrecy data rate and the payment for the interference to cellular users. The game strategy of cellular users depends on the price factor α, and the game strategy of D2D users depends on the transmission power P_D. The Stackelberg game is formed to maximize the cellular utility function in the first stage and then the utility function of the D2D user in the second stage.

Power control is a commonly used mechanism for improving network EE. In Bernardo and De Leon [2016], the authors studied the trade-off between PLS and EE of massive MIMO in an HetNet. An optimization model is presented to minimize the total power consumption of the network while satisfying the security level against eavesdroppers by assuming that the BS has imperfect channel knowledge of the eavesdroppers. The simulation results indicate that a highly dense network topology is an effective solution for achieving high capacity, cellular EE, and secure, reliable communication channels.

- Relay: As shown in Figure 3.11, cooperation with the relay can be used to help the sender to secure the signal transmission. In Nguyen et al. [2016], two relay selection protocols, namely optimal relay selection (ORS) and partial relay selection (PRS), are proposed to secure an energy-harvesting relay system in 5G wireless networks. The system model is shown in Figure 3.11, which consists of multiple relay nodes and assumes there is no direct link between sender and receiver. The power beacon is equipped with multiple antennas, which can be used to strengthen the energy harvested. The ORS chooses the aiding relay to maximize the secrecy capacity of the system by assuming that the source has full knowledge of channel state information (CSI) on each link. The PRS selects the helping relay based on partial CSI. The system is composed of a high-powered beacon equipped with multiple antennas, several relays, a destination node, and an eavesdropper with a single antenna. Two energy-harvesting scenarios that aim to maximize energy harvesting for the source and selected relay are investigated. The analytical and asymptotic expressions of secrecy outage probability for both relay selection protocols are presented. The numerical results show that ORS can significantly enhance the security of the proposed system model and can achieve full secrecy diversity order, while PRS can only achieve unit secrecy diversity order regardless of the energy harvest strategies. PRS that maximizes energy harvesting for relay strategy has a better secrecy performance than the one based on maximizing energy harvesting for source. Moreover, the results show that the secrecy performance of the considered system is impacted significantly by the duration of the energy harvest process.

To tackle the complex issue of relay selection in 5G large-scale secure two-way relay amplify-and-forward (TWR-AF) systems with massive relays and eavesdroppers, in Zhang et al. [2017b], the authors proposed a distributed relay selection criterion that does not require the information of sources signal-to-noise ratio (SNR), channel estimation, or the knowledge of relay eavesdropper links. The proposed relay selection is made based on the received power of relays and knowledge of the average channel information between the source and the eavesdropper. The system model includes two source nodes, a number of legitimate relay nodes, and multiple passive eavesdroppers. Each node has a single antenna. The cooperation of eavesdroppers is considered. In TWR-AF, the eavesdropper receives overlapping signals from the two sources in each time slot, with one of the signals acting as jamming noise. The analytical results show that the number of eavesdroppers has a severe impact on the secrecy performance. The simulation results show that the performance of the proposed low-complexity criterion is very close to that of the optimal selection counterpart.

Considering eavesdroppers and relay with both single and multiple antennas, in Xu et al. [2016b], the transmission design for secure relay communications in 5G networks is studied by assuming no knowledge of the number or the locations of eavesdroppers. The locations of eavesdroppers are generated by a homogeneous Poisson Point Process. A randomize-and-forward relay strategy is proposed to secure multi-hop communications. The secrecy outage probability of the two-hop transmission is derived. A secrecy rate maximization problem is formulated with a secrecy outage probability constraint. It gives the optimal power allocation and secrecy rate. Simulation results show that the secrecy outage probability can be improved by equipping each relay with multiple antennas. The secrecy throughput is enhanced, and secure coverage is extended by appropriately using relaying strategies.

- Artificial noise: Artificial noise can be introduced to secure the intended signal transmission. In a heterogeneous cellular network with randomly located eavesdroppers, [Wang et al., 2016b] proposed an association policy that employs an access threshold for each user to maximize the truncated average received signal power beyond the threshold in artificial-noise-aided multi-antenna secure transmission under a stochastic geometry framework. The tractable expression of connection probability and secrecy probability for a randomly located legitimate user is investigated. The network secrecy throughput and minimum secrecy throughput of each user are presented under the constraints of connection and secrecy probabilities. The accuracy of the analysis is verified through numerical results.

Assuming that the sender is armed with multiple antennas, in Ju et al. [2016], an artificial noise transmission strategy is proposed to secure the transmission against an eavesdropper with a single antenna in millimeter-wave systems. A millimeter-wave channel is modeled with a ray cluster-based spatial channel model. The sender has partial CSI knowledge of the eavesdropper. The proposed transmission strategy depends on the directions of the destination and the propagation paths of the eavesdropper. The secrecy outage probability is used to analyze the transmission scheme. An optimization problem based on minimizing the secrecy outage probability with a secrecy rate constraint is presented. To solve the optimization problem, a closed-form optimal power allocation between the information signal and artificial noise is derived. The secrecy performance of the millimeter-wave system is significantly influenced by the relationship between the propagation paths of the destination and the eavesdropper. The numerical results show that the secrecy outage mostly occurs if the common paths are large or the eavesdropper is close to the transmitter.

To improve EE of the security method using artificial noise, in Zappone et al. [2016], an optimization problem is formulated to maximize the secrecy EE by assuming imperfect CSI of eavesdropper at the transmitter. The system is modeled with one legitimate transmitter with multiple antennas, one legitimate receiver, and one eavesdropper, each with a single antenna. Artificial noise is used at the transmitter. Resource allocation algorithms are used to solve the optimization problem with a correlation between transmit antennas.

A method that employs fractional programming and sequential convex optimization is proposed to compute first-order optimal solutions with polynomial complexity.

- Signal processing: Besides the three methods above to provide data confidentiality, in Chen et al. [2016a], the authors proposed an original symbol phase rotated (OSPR) secure transmission scheme to defend against eavesdroppers armed with an unlimited number of antennas in a single cell. Perfect CSI and perfect channel estimation are assumed. The BS randomly rotates the phase of original symbols before they are sent to legitimate user terminals. The eavesdropper cannot intercept signals; only the legitimate users are able to infer the correct phase rotations and recover the original symbols. The symbol error rate of the eavesdropper is studied, which proves that the eavesdropper cannot intercept the signal properly as long as the BS is equipped with a sufficient number of antennas. Considering multiple eavesdroppers in Qin et al. [2016], the authors analyzed the secure performance on a large-scale downlink system using non-orthogonal multiple access (NOMA). The system considered contains one BS, M NOMA users, and eavesdroppers randomly deployed in a finite zone. A protected zone around the source node is adopted to enhance the security of the random network. Channel statistics for legitimate receivers and eavesdroppers and secrecy outage probability are presented. The user pair technique is adopted among the NOMA users. Analytical results show that the secrecy outage probability of NOMA pairs is determined by the NOMA users with poorer channel conditions. Simulation results show that secrecy outage probability decreases when the radius of the protected zone increases, and secrecy outage probability can be improved by reducing the scope of the user zone as the path loss decreases. In Xu et al. [2016a], the authors proposed a dynamic coordinated multipoint transmission (CoMP) scheme for BS selection to enhance secure coverage. Considering co-channel interference and eavesdroppers, an analysis of the secure coverage probability is presented. Both analytical and simulation results show that by utilizing CoMP with a proper BS selection threshold, the secure coverage performance can be improved, while secure coverage probability decreases with excessive cooperation. The proposed CoMP scheme has a better performance in resisting more eavesdroppers than the no-CoMP scheme. In Deng et al. [2015], massive MIMO is applied to HetNets to secure data confidentiality in the presence of multiple eavesdroppers. Tractable upper-bound expressions for the secrecy outage probability of HetNet users are presented, revealing the considerable gains in secrecy performance achievable through the use of massive MIMO. Moreover, the paper examines the connection between the density of the pico cell BS and the secrecy outage probability of HetNet users.
- Cryptographic methods: In addition to the PLS solutions discussed earlier, cryptographic techniques are employed to ensure data confidentiality through the use of secret keys for data encryption. Public-key cryptography offers a viable solution for the distribution of symmetric keys. To reduce the computational cost of encryption, symmetric-key cryptography is adopted for data encryption. In Eiza et al. [2016], a participating vehicle can send its random symmetric key, which is encrypted using TA's public key. The symmetric key is used to encrypt the message between TA, DMV, and participating vehicles. A one-time encryption key is also encrypted by a public key. The one-time encryption key is used to encrypt the video. In Zhang et al. [2017a], an initial symmetric session key is

negotiated between the client and a physician after establishing the client/server relationship. The symmetric key is then used for the data transmission between the client and the physician.

3.5 Key Management

Key management refers to the process of securely creating, storing, distributing, and revoking cryptographic keys used in a cryptographic system. Cryptographic keys are used to encrypt and decrypt information, verify digital signatures, and provide authentication in various applications such as secure communication, data storage, and online transactions.

In Sedidi and Kumar [2016], three novel key exchange protocols, which have different levels of computational time, computational complexity, and security, for D2D communications are proposed based on the Diffie–Hellman (DH) scheme. Details of the key exchange schemes are shown in Figure 3.13. The threat analysis of all three proposed protocols under common brute force and man-in-the-middle (MITM) attacks is presented. The authors have conducted a performance analysis of the proposed protocols to assess the effectiveness of their security services, including confidentiality, integrity, authentication, and non-repudiation. This analysis is based on a combination of theoretical evaluation and practical experimentation. The performance analysis confirms the feasibility of the proposed protocols, demonstrating that they can be implemented with reasonable levels of communication overhead and computational time.

For D2D group communications use cases, in Abd-Elrahman et al. [2015], a group key management (GKM) mechanism is proposed to secure the exchanged D2D message during

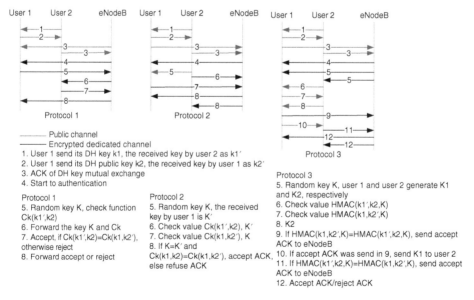

Figure 3.13 Three key exchange schemes in Sedidi and Kumar [2016]. Source: Adapted from Sedidi and Kumar [2016].

the discovery and communication phases. There are five security requirements in the proposed GKM, namely forward secrecy (users that have left the group should not have access to the future key), backward secrecy (new users joining the session should not have access to the old key), collusion freedom (fraudulent users could not deduce the current traffic encryption), key independence (keys in one group should not be able to discover keys in another group), and trust relationship (do not reveal the keys to any other part in the same domain or any part in a different domain). ID-based cryptography (IBC) scheme based on Elliptic Curve Cryptography (ECC) for securing multicast group communications is presented. The steps of the proposed protocol include secret key generation, elliptic curve digital signature algorithm, signature verification, group formation procedure, key generation, join process, and leave process. The master key and private key generations are based on IBC and ECC schemes. The overhead for communications, re-keying messages, and key storage is assessed. The weakness of the IBC scheme and the ways of creating and using GKM are compared. The overall performance comparisons show that the proposed GKM has an enhancement in both the protocol complexity and security level compared with other works.

ECC is also adopted for the proposed LRSA protocol in Zhang et al. [2017a]. The network manager generates a partially private and partially public key for the client and the physician after the registration. And once the client and the physician establish the client/server relationship, an initial systematic session key can be set up for the data transmission.

3.6 Privacy

As previously discussed, the emergence of 5G wireless networks and their support for vertical industries such as m-health care and intelligent transportation raises significant concerns about the potential for privacy breaches [Huawei, 2015]. The data flows in 5G wireless networks carry extensive personal privacy information such as identity, position, and private contents. In some cases, privacy leakage may cause serious consequences. Depending on the privacy requirements of the applications, privacy protection is a big challenge in 5G wireless networks. There has already been research work considering location privacy and identity privacy.

Regarding location privacy, in Farhang et al. [2015], to protect the location and preferences of users that can be revealed with associated algorithms in HetNets, a decentralized algorithm for access point selection is proposed based on a matching game framework, which is established to measure the preferences of mobile users and BSs with physical layer system parameters. A differentially private Gale–Shapley matching algorithm is developed based on differential privacy. Utilities of mobile users and access points are proposed based on packet success rate. Simulation results show that the differentially private algorithm can protect location privacy with a good service quality based on the mobile users' utility. In Ulltveit-Moe et al. [2011], a location-aware mobile intrusion prevention system (mIPS) architecture with privacy enhancement is proposed. The authors presented the mIPS requirements and possible privacy leakage from managed security services.

In Zhang et al. [2017a], contextual privacy is defined as the privacy of data source and destination. The identity of the source client is encrypted by a pseudo-identity of the source client with the public key of the physician using certificateless encryption mode. Meanwhile, the identity of the intended physician is also encrypted with the public key of the network manager. Through these two encryption steps, contextual privacy can be achieved. For the proposed reporting service in Eiza et al. [2016], privacy is an essential requirement to gain the acceptance and participation of people. The identity and location information of a vehicle should be preserved against illegal tracing. Meanwhile, a reporting vehicle should be able to reveal its identity to the authorities for special circumstances. The pseudonymous authentication schemes are applied to achieve conditional anonymity and privacy.

3.7 Conclusion

In this chapter, current security solutions for 5G networks were surveyed based on different security services such as authentication, availability, data confidentiality, key management, and privacy. The security requirements of 5G wireless networks are evaluated with respect to each security service, taking into account the specific needs and challenges of the new network environment. To meet the performance demands of 5G wireless network services, security and privacy mechanisms must be efficient, flexible, and adaptable.

4

An Efficient Security Solution Based on Physical Layer Security in 5G Wireless Networks

The use of HetNets is widely considered a promising approach for improving wireless coverage and throughput in 5G networks. Due to its heterogeneous characteristics, HetNet offers several benefits, including higher capacity, wider coverage, and improved energy and spectrum efficiency. However, HetNet also brings complex interference management to the network. There is a lot of work on interference management in HetNet. Many methods are based on reducing interference, which can increase the eavesdropping probability in the network. The focus of this chapter is to present a security analysis of a physical layer security solution that not only addresses interference management but also provides protection against eavesdropping attacks [Fang et al., 2017b, 2019a]. The proposed interference management mechanism aims to maximize the secrecy rate of a cellular user under an eavesdropping attack.

4.1 Enhancing 5G Security Through Artificial Noise and Interference Utilization

This section provides an overview of research on physical layer security, which includes the utilization of artificial noise and existing interference.

Adding Artificial noises is one of the popular methods to improve the physical layer security in HetNet. In Wang et al. [2017], the authors studied physical layer security in multi-antenna small-cell networks with eavesdroppers randomly distributed. The artificial noise-aided transmission at each base station is adopted to improve the secrecy rate performance. The introduced artificial noise can increase the inter-cell interference in a multi-cell cellular network degrading the reliability of the network and the secrecy rate performance of users. In Wang et al. [2016a], the impact of artificial noise on the network reliability and security of users was investigated. The analytical results show that artificial noise can be an efficient solution for securing communications in a multi-cell cellular network with an optimized power allocation between the desired signals and artificial noise. Artificial noise can be applied to secure communications in the multiple-input-and multiple-output (MIMO) channel. Channel estimation can provide a source of randomness which can be utilized to generate secret keys for the sender and receiver in MIMO wireless channels. By utilizing MIMO wireless channels and artificial noise, in Harper and Ma [2016], encrypted symbols were used as messages for the intended receiver and as noise for any potential eavesdroppers. Both secrecy rate and power consumption are improved by the proposed method compared to the standard artificial noise scheme. In Zhao et al. [2016, 2018], an artificial noise scheme was proposed to disrupt the external eavesdropping attack without introducing any additional interference to the legitimate network. In Özçelikkale and Duman [2015], an artificial noise-aided scheme where transmitters can broadcast noise in addition to data to confuse eavesdroppers was proposed. An optimization problem was formulated to minimize the total mean-square error of the legitimate receivers and keep the error values of the eavesdroppers above a target level. Extra power is needed for all these artificial noises based physical layer security methods. Moreover, the separation of interference management and security management in HetNet makes the system inefficient.

Instead of introducing artificial noise into the network, there is research work utilizing existing interference of the network to provide security for the users. In particular, secrecy capacity is defined and considered by utilizing the advantage of the interference caused by D2D communications as a possible help against eavesdroppers [Yue et al., 2013]. The secrecy outage probability is applied to better depict the imperfect channel state information at the eavesdropper. The simulation results of secrecy outage probability show that the interference caused by D2D communications can help guarantee the secrecy demand of the cellular user and increase the network throughput as well as spectrum efficiency (SE) and energy efficiency (EE). In Sheikholeslami et al. [2012], the authors proposed schemes where multiple transmitters cooperatively send their signals to confuse the eavesdroppers. The aggregated interference by all the transmitters is utilized to protect the signal transmitted by a specific transmitter. In Sibomana et al. [2015], the physical layer security of the secondary user was investigated considering the interference from primary users over cognitive radio networks. The numerical results show that interference from primary users can improve the security of secondary users. Besides considering the security of secondary users, in Zhang et al. [2016], interference from secondary users was used to improve the secrecy capacity of the primary users. A tradeoff between the channel capacity of the secondary users and the secrecy capacity of the primary users was studied. In Cao et al. [2018a], two schemes were proposed to improve the sum rate of secondary users while guaranteeing the secrecy rate of primary users for cognitive radio applications. The simulation results show that the proposed schemes can effectively secure cognitive radio networks.

4.2 A HetNet System Model and Security Analysis

4.2.1 System Model and Threat Model

A two-tier HetNet system model is shown in Figure 4.1. In this studied system model, there is one macro-base station (MBS), a number of I small base stations (SBSs), an eavesdropper, and multiple users. Assume that the eavesdropper is eavesdropping on a user associated with the MBS in the first tier. The MBS and SBSs are equipped with multiple antennas for applying the massive MIMO technique. A single antenna is applied to each user and the eavesdropper. Let the number of antennas of the MBS and each SBS be A_m and A_s, respectively. Let the maximum number of users associated with the MBS and each SBS be M and N, respectively, where $A_m > M$ and $A_s > N$.

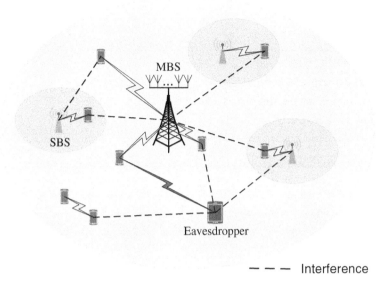

MBS

SBS

Eavesdropper

— — — Interference

Figure 4.1 The studied two-tier HetNet system model.

The MBS is located in the center of the coverage area with transmit power P_m as shown in Figure 4.1 in the first tier. D2D communications are applied with transmit power P_d in the first tier using the same spectrum as the downlink of the MBS. Let the number of D2D pairs be D and the signal QoS requirement of users associated with the MBS be γ_m, where $m \in [1, M]$. Define the signal and channel vector from the MBS to the mth user associated with the MBS as s_m and h_m, respectively. The precoding vector for the mth user associated with the MBS is w_m. Beamforming is adopted in base stations.

The second tier is within the coverage of the MBS. The SBSs are distributed following an independent Poisson point process Φ_s with density λ_s. Define the transmit power of each SBS as P_s. This study does not take into account any interference that may be caused by D2D communications to users associated with SBSs. Let the signal QoS requirement of the nth user associated with the ith SBSs be γ_{in}, where $i \in [1, I]$, $n \in [1, N]$. Table 4.1 lists all variables used in this study.

Since massive MIMO is used in all SBSs, the interference from SBSs to the MBS can be utilized to improve the secrecy rate of users associated with the MBS. Moreover, some SBSs can reuse part of the spectrum used by the MBS. For these SBSs with a short distance from each other, an orthogonal spectrum can be applied to eliminate the interference between SBSs. The interference between SBSs and D2D users to users associated with SBS are ignored since the transmit power of D2D users is small. As shown in Figure 4.1, for users associated with SBSs that reuse spectrum with the MBS, interference from MBS is considered. Similar to users associated with the MBS, the interference from spectrum reuse SBSs and D2D communications is considered.

This study focuses on eavesdropping as one of the types of adversaries in the system. Due to the signal strength, the user associated with the MBS is easier to be eavesdropped. The assumption is that the eavesdropper is attempting to intercept the communication between

Table 4.1 Definitions of variables.

Variable	Definition
s_m	The signal from MBS to the mth user associated with
s_{in}	The signal from the ith SBS to the nth user associated with the ith SBS
h_m	The channel vector from the MBS to the mth user associated with MBS
h_{im}	The channel vector from the ith SBS to the mth user associated with MBS
h_e	The channel vector from the MBS to the eavesdropper
h_{ie}	The channel vector from the ith SBS to the eavesdropper
$h_{M,in}$	The channel vector from the MBS to the nth user associated with the ith SBS
h_{in}	The channel vector from the ith SBS to the nth user associated with the ith SBS
w_m	The precoding vector for the mth user associated with MBS
w_{in}	The precoding vector for the nth user associated with the ith

a legitimate user and the MBS. Different interference is considered in this study. As shown in Figure 4.1, the eavesdrop has interference from D2D users, users associated with the MBS, and SBSs, which share the same spectrum with the MBS.

4.2.2 Security Analysis

This subsection examines the interference that occurs in the HetNet system, analyzing the received signals and SINR of both legitimate users and any eavesdroppers present.

The interference in the studied HetNet system includes interference from MBS, SBSs, and D2D communications. With the distance limitation of wireless communications, the interference from D2D communications is only considered to users associated with the MBS and the eavesdropper under a certain distance threshold θ. Define the D2D interference indicator as l_m^d for the mth user associated with the MBS and with the dth D2D user pair,

$$l_m^d = \begin{cases} 1, & L_m^d \leq \theta \\ 0, & otherwise \end{cases} \tag{4.1}$$

where L_m^d represents the distance between the mth user associated with the MBS and the dth D2D user pair. Similar to the D2D interference indicator, the interference indicator of eavesdroppers from D2D users l_e^d can be defined. This study assumes that the MBS controls D2D communications under the SINR requirement of users associated with the MBS. If a D2D pair communication degrades a legitimate user in the first tier unqualified based on the SINR requirement, the D2D pair communication will not be approved. Otherwise, the MBS will approve the D2D pair communication. The D2D pair communication may degrade the eavesdropper link based on the distance. Therefore, to improve the secrecy rate of the user under an eavesdropping attack, assume that the MBS only approves the D2D pairs near the possible eavesdropper, which satisfy $l_e^d = 1$. Let the interference from D2D user under the distance threshold be f_{D2D},

$$f_{D2D} = P_d h_d^{D2D} \|\theta\|^{-\alpha}, \tag{4.2}$$

where h_d^{D2D} is the fading factor, α is the path-loss exponent, and $\alpha \geq 2$.

Besides the interference from D2D communications, there is also interference due to spectrum reuse. For users associated with the MBS, each of them may have interference from other users in the MBS, users in the SBSs with the same spectrum as the MBS, and D2D communications. Gaussian noise is also considered. For the mth user associated with the MBS, the interference signal at the mth user associated with the MBS is given by

$$f_m = \sum_{m' \neq m}^{M} h_m w_{m'} s_{m'} + \sum_{i=1}^{I} \sum_{n=1}^{N} h_{im} w_{in} s_{in} l_i + \sum_{d=1}^{D} f_{D2D} l_m^d + n_m, \tag{4.3}$$

The first term is the interference from the MBS due to the signal to other users in the first tier. The second term indicates the interference signal from SBSs, which use the same spectrum as the MBS. The l_i is the indicator of the spectrum allocation. If $l_i = 1$, which presents the ith SBS allocated the same spectrum as the MBS. To improve the security of the cellular user, the SBSs next to the eavesdropper will be assigned to reuse the spectrum with the MBS. The third term represents the interference signal from D2D users, and the last term is the Gaussian noise following independent and identically distributed $N(0, \sigma_m^2)$. Therefore, the received signal of the mth user associated with the MBS is

$$y_m = h_m w_m s_m + f_m, \tag{4.4}$$

where the first term represents the desired signal from the MBS. Based on the number of antennas at the MBS and SBSs, where $h_m \in C^{1 \times A_m}$, $h_{im} \in C^{1 \times A_s}$, $w_m \in C^{A_m \times 1}$, and $w_{in} \in C^{A_s \times 1}$. The precoding vectors for the users associated with the MBS should satisfy the power constraint as $\sum_{m=1}^{M} \|w_m\|^2 = P_m$. Similarly, the precoding vectors for the users associated with SBS should satisfy the power constraint as $\sum_{n=1}^{N} \|w_{in}\|^2 = P_s$. Therefore, for the mth user associated with the MBS, the SINR is

$$SINR_m = \frac{|h_m w_m|^2}{IN_m} \tag{4.5}$$

where

$$IN_m = \sum_{m' \neq m}^{M} |h_m w_{m'}|^2 + \sum_{i=1}^{I} \sum_{n=1}^{N} |h_{im} w_{in} l_i|^2 + \sum_{d=1}^{D} |f_{D2D} l_m^d|^2 + \sigma_m^2 \tag{4.6}$$

IN_m is the sum of interference and noise power of the mth user.

For the eavesdropper, the interference signal is similar to users associated with the MBS. Gaussian noise is also considered. To simplify the notations, assume that the user under an eavesdropping attack is the user number 1. Then the interference signal at the eavesdropper is

$$f_e = \sum_{m=2}^{M} h_e w_m s_m + \sum_{i=1}^{I} \sum_{n=1}^{N} h_{ie} w_{in} s_{in} l_i + \sum_{d=1}^{D} f_{D2D} l_e^d + n_e, \tag{4.7}$$

The first term indicates the interference signal from the MBS due to the signal to other users in the first tier. The second term is the interference signal from SBSs, which share the same spectrum as the MBS. The third term represents the interference signal from D2D users, and the last term is the Gaussian noise following independent and identically distributed

$N(0, \sigma_e^2)$, based on the number of antennas at the MBS and SBSs, where $h_e \in C^{1 \times A_m}$, $h_{ie} \in C^{1 \times A_s}$. Therefore, the received signal from the eavesdropper is

$$y_e = h_e w_1 s_1 + f_e, \tag{4.8}$$

where the first term represents the desired eavesdropped signal from the MBS. The SINR of the eavesdropper can be calculated as follows.

$$SINR_e = \frac{|h_e w_1|^2}{IN_e} \tag{4.9}$$

where

$$IN_e = \sum_{m=2}^{M} |h_e w_m|^2 + \sum_{i=1}^{I} \sum_{n=1}^{N} |h_{ie} w_{in} l_i|^2 + \sum_{d=1}^{D} |f_{D2D} l_e^d|^2 + \sigma_e^2 \tag{4.10}$$

IN_e is the sum of interference and noise power of the eavesdropper.

For users associated with SBSs, the interference can be caused by SBSs and the MBS. Due to the limited transmit power of D2D communications, interference from D2D communications is not considered. The interference signal at the nth user associated with the ith SBS is

$$f_{in} = \sum_{n' \neq n}^{N} h_{in} w_{in'} s_{in'} + \sum_{m=1}^{M} h_{M,in} w_m s_m l_i + n_{in} \tag{4.11}$$

The first term is the interference signal from the ith SBS. The second term indicates the interference signal from the MBS, and the last term is the Gaussian noise following independent and identically distributed $N(0, \sigma_s^2)$, where $h_{M,in} \in C^{1 \times A_m}$. Therefore, the signal received at the nth user associated with the ith SBS is

$$y_{in} = h_{in} w_{in} s_{in} + f_{in} \tag{4.12}$$

where the first term represents the desired signal from the nth SBS. The SINR of the nth user associated with the ith SBS is written as

$$SINR_{in} = \frac{|h_{in} w_{in}|^2}{IN_{in}} \tag{4.13}$$

where

$$IN_{in} = \sum_{n' \neq n}^{N} |h_{in} w_{in'}|^2 + \sum_{m=1}^{M} |h_{M,in} w_m l_i|^2 + \sigma_{in}^2 \tag{4.14}$$

Based on the above preliminary analysis, a security analysis of the proposed interference management mechanism is presented in Section 4.3 to maximize the secrecy rate of the user under an eavesdropping attack with constraints of SINR requirements of other users.

4.3 Problem Formulation and Analysis

This section provides a security analysis of the proposed interference management mechanism, taking into account the spectrum reuse between some SBSs and MBS, as well as transmit beamforming and D2D communications. A secrecy rate optimization problem of the user under an eavesdropping attack is formulated. The interference from SBSs and D2D communications to the user associated with the MBS is considered.

4.3.1 Maximum Secrecy Rate

The secrecy rate optimization problem of the user under an eavesdropping attack is formulated as

$$\max_{w_m, w_{in}, l_i, l_m^d, l_e^d} \log(1 + SINR_1) - \log(1 + SINR_e)$$

$$\text{s. t.} \sum_{m=1}^{M} \|w_m\|^2 \leq P_m \tag{4.15}$$

$$\sum_{n=1}^{N} \|w_{in}\|^2 \leq P_s, i \in [1, I],$$

$$SINR_m \geq \gamma_m, m \in [2, M],$$

$$SINR_{in} \geq \gamma_{in}, i \in [1, I], n \in [1, N].$$

The first two constraints are based on the transmit power limitation of the MBS and SBSs. The last two constraints are based on the SINR requirements of users associated with MBS and SBSs. In this optimization problem, transmit beamforming, spectrum reuse, and D2D communications are considered for utilizing the interference to maximize the secrecy rate of the eavesdropped user. w_m and w_{in} are the precoding vectors for the transmit beamforming. l_i depends on the spectrum allocation between SBSs and the MBS. l_m^d and l_e^d are the indicators of interference of D2D communications based on distance from D2D pairs to users in first tier and to the eavesdropper.

4.3.2 The Proposed Algorithm

Since the logarithm function is an increasing function, the optimization problem in Eq. (4.15) is equivalent to the optimization problem as

$$\max_{w_m, w_{in}, l_i, l_m^d, l_e^d} \frac{(1 + SINR_1)}{(1 + SINR_e)} \tag{4.16}$$

with the same constraints in Eq. (4.15). The optimization problem is not convex. With these five parameters, it cannot be solved based on linear optimization methods. Based on the previous analysis, l_i, l_m^d, and l_e^d can affect the SINR of legitimate users and the eavesdropper. Therefore, the spectrum allocation and D2D communication indicators can be predetermined based on the distance between base stations and D2D pairs to the eavesdropper and the user.

Let $W_m = w_m \times w_m^T$, $W_{in} = w_{in} \times w_{in}^T$, $H_m = h_m^T \times h_m$, $H_{im} = h_{im}^T \times h_{im}$, $H_e = h_e^T \times h_e$, $H_{ie} = h_{ie}^T \times h_{ie}$, $H_{in} = h_{in}^T \times h_{in}$, and $H_{M,in} = h_{M,in}^T \times h_{M,in}$. Without loss of generality, assume that $\sigma_m^2 = \sigma_e^2 = \sigma_{in}^2 = 1$. Based on the SINR information presented in the preliminary analysis in Section 4.3.1, Eq. (4.16) and all the constraints can be rewritten as

$$\max_{w_m, w_{in}, l_i, l_m^d, l_e^d} \frac{1 + \frac{Tr(H_1 \times W_1)}{IN_1'}}{1 + \frac{TR(H_e \times W_1)}{IN_e'}}$$

$$\text{s. t.} \sum_{m=1}^{M} Tr(W_m) \leq P_m, \tag{4.17}$$

$$\sum_{n=1}^{N} Tr(W_{in}) \leq P_s, i \in [1, I],$$

$$\frac{Tr(H_m \times W_m)}{IN'_m} \geq \gamma_m, m \in [2, M],$$

$$\frac{Tr(H_{in} \times W_{in})}{IN'_{in}} \geq \gamma_{in}, i \in [1, I], n \in [1, N].$$

where

$$IN'_m = \sum_{m' \neq m}^{M} Tr\left(H_m \times W_{m'}\right) + \sum_{i=1}^{I}\sum_{n=1}^{N} Tr\left(H_{im} \times W_{in} \times l_i\right) + \sum_{d=1}^{D} |f_{D2D} l_m^d|^2 + 1,$$

$$IN'_{in} = \sum_{n' \neq n}^{N} Tr(H_{in} \times W_{in'}) + \sum_{m=1}^{M} Tr(H_{M,in} \times W_m \times l_i) + 1. \qquad (4.18)$$

The $Tr()$ represents the trace of a square matrix. The optimization problem in Eq. (4.17) is not convex. To convert it into a convex optimization problem, first determine the spectrum allocation and D2D communication indicators as l_i, l_m^d, and l_e^d based on the location of base stations and D2D pairs to the eavesdropper and the user. Therefore, the problem can be simplified as follows:

$$\max_{w_m, w_{in}} \frac{1 + \frac{Tr(H_1 \times W_1)}{IN'_1}}{1 + \frac{TR(H_e \times W_1)}{IN'_e}} \qquad (4.19)$$

with same constraints in Eq. (4.17). Then a slack variable $\xi = SINR_e$ is introduced to convert Eq. (4.19) into a convex problem. Since ξ is also related to w_m and w_{in}, then the optimization problem in Eq. (4.19) can be equivalently transformed into the following optimization problem:

$$\max_{w_m, w_{in}} \frac{1 + \frac{Tr(H_1 \times W_1)}{IN'_1}}{1 + \xi}$$

$$\text{s. t.} \frac{Tr(H_e \times W_1)}{IN'_e} \leq \xi \leq Tr(H_1) P_m \qquad (4.20)$$

The constraint of ξ represents the lower and upper bounds. For a fixed value of ξ, a corresponding value of the following optimization problem can be formed:

$$\max_{W_m, W_{in}} \frac{Tr(H_1 \times W_1)}{IN'_1} \qquad (4.21)$$

with the same constraints in Eq. (4.17). Based on the values of ξ and the corresponding value of $\frac{Tr(H_1 \times W_1)}{IN'_1}$, the one-dimensional line search method can be applied to solve the problem in Eq. (4.17).

Since Eq. (4.21) is a linear-fractional program which can be transformed into a linear program by the Charnes–Cooper transformation with

$$Y_m = \frac{W_m}{IN'_1}, Y_{in} = \frac{W_{in}}{IN'_1}, \zeta = \frac{1}{IN'_1} \qquad (4.22)$$

Then Eq. (4.21) can be transformed into a semi-definite programming duality (SDD) problem as

$$\max_{Y_m, Y_{in}} Tr(H_1 \times Y_1)$$

$$\text{s. t.} \sum_{m=1}^{M} Tr(Y_m) \leq P_m \zeta,$$

$$\sum_{n=1}^{N} Tr(Y_{in}) \leq P_s \zeta, i \in [1, I],$$

$$Tr(H_m \times W_m) \geq \gamma_m IN''_m, m \in [2, M],$$

$$Tr(H_{in} \times Y_{in}) \geq \gamma_{in} IN''_{in}, i \in [1, I], n \in [1, N],$$

$$Tr(H_e \times Y_1) \leq \xi IN''_e,$$

$$\sum_{m=2}^{M} Tr(H_m \times Y_m) + \sum_{i=1}^{I} \sum_{n=1}^{N} Tr(H_{i1} \times Y_{in} \times l_i) + \left(\sum_{d=1}^{D} |f_{D2D} l_m^d|^2 + 1 \right) \zeta = 1,$$

$$Y_m, Y_{in} \geq 0, \zeta \geq 0. \tag{4.23}$$

where

$$IN''_m = \sum_{m' \neq m}^{M} Tr(H_m \times Y_{m'}) + \sum_{i=1}^{I} \sum_{n=1}^{N} Tr(H_{im} \times Y_{in} \times l_i) + \left(\sum_{d=1}^{D} |f_{D2D} l_m^d|^2 + 1 \right) \zeta,$$

$$IN''_{in} = \sum_{n' \neq n}^{N} Tr(H_{in} \times Y_{in'}) + \sum_{m=1}^{M} Tr(H_{M,in} \times Y_m \times l_i) + \zeta,$$

$$IN''_e = \sum_{m=2}^{M} Tr(H_e \times Y_m) + \sum_{i=1}^{I} \sum_{n=1}^{N} Tr(H_{ie} \times Y_{in} \times l_i) + \left(\sum_{d=1}^{D} |f_{D2D} l_e^d|^2 + 1 \right) \zeta. \tag{4.24}$$

The optimization solution of Eq. (4.23) is denoted by $(Y_m^*, Y_{in}^*, \zeta^*)$. Based on Eq. (4.22), the optimal solution of Eq. (4.21) can be obtained as $W_m^* = \frac{Y_m^*}{\zeta^*}$ and $W_{in}^* = \frac{Y_{in}^*}{\zeta^*}$. With the lower s and upper bound of ξ, the optimal solution ξ^* can be obtained by applying a one-dimensional line search method, such as the Fibonacci search. The optimization problem in Eq. (4.23) can be solved by CVX [Guimaraes et al., 2015]. The flow chart of Algorithm 4.1 is shown in Figure 4.2 [Fang et al., 2017b].

Algorithm 4.1 The proposed method

---

1) State-related parameters of the channel, BS, spectrum reuse, and D2D communication parameters and one-dimensional search method.
2) Calculate the upper bound of the slack variable.
3) Golden section search of the slack variable and find the according value of Eq. (4.21).
4) Obtain the optimal solution. End.

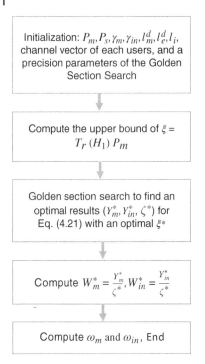

Figure 4.2 Flow chart of Algorithm 4.1.

4.4 Numerical and Simulation Results

This section presents the numerical results and discussions. The simulation is completed using Matlab. First, evaluate the security of the user's information transmission in the presence of eavesdropping, by comparing the achieved secrecy rate using the proposed interference management mechanism against three alternative methods: complete interference elimination through non-reused spectrum allocation without beamforming, interference management using only D2D communication without beamforming, and interference management using only spectrum reuse between MBS and SBSs without beamforming. The secrecy rate performance of a downlink HetNet cellular user is analyzed. In the simulation, the system model includes one MBS and two SBSs with spectrum reuse as the MBS. In the first tier, there are two users associated with the MBS. In each SBS, there is one user. D2D communications can be established at various distances from the eavesdropper's location. Assuming 10 antennas for the MBS and four antennas for each SBS, the MBS's transmit power is set to 10 W. The SINR threshold for all users is set to 0.6.

The secrecy rate performance of the proposed method and three other interference management techniques are compared in Figure 4.3. The label "Non" refers to total interference elimination through non-reuse of spectrum without beamforming. "Only D2D" implies that only D2D interference is taken into account. "Only SR" denotes that only spectrum reuse between MBS and SBSs is considered. The secrecy rate is independent of the transmit power of SBSs for both the "Non" and "Only D2D" methods. In this simulation, the first tier is assumed to involve D2D communications, which may impact both the cellular user and the eavesdropper. Figure 4.3 represents 5 D2D pairs. The label "The proposed" refers to the proposed methods for beamforming, spectrum reuse,

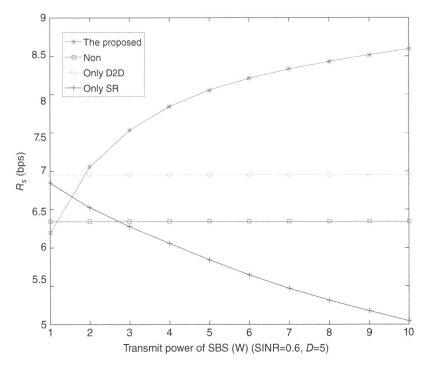

Figure 4.3 Secrecy rate of the user under eavesdropping attack.

and interference management in D2D communications. Without beamforming, D2D interference can improve the secrecy rate at a certain level. The use of spectrum reuse is the only method that can result in a decrease in secrecy rate as the transmission power of SBSs increases, without the use of beamforming. When beamforming is not employed, the use of spectrum reuse may lead to greater interference for the legitimate user compared to the eavesdropper.

The secrecy rate of a user facing an eavesdropping attack was examined for different numbers of D2D pairs in Figure 4.4. The simulation results indicate that the secrecy rate is maximized with four D2D pairs. This is because the presence of an additional D2D pair compared to three pairs can degrade the eavesdropper's communication more than the legitimate user. However, for five pairs of D2D users, one pair may degrade the legitimate user's communication more than the eavesdropper's. Therefore, not all the D2D interference in the network can be utilized to improve the secrecy rate. The proposed scheme's performance can be further improved by incorporating an iteration method into the D2D pair selection process.

Figure 4.5 illustrates the secrecy rate performance considering different SINR requirements. For the simulation, all users were assigned the same SINR requirement. As higher SINR requirements limit the possibility of utilizing interference, the results show that a lower SINR requirement leads to a higher secrecy rate. In typical cellular communication systems, the SINR requirement is set to 0.6. When compared to a SINR requirement of 0.7, a 0.6 SINR requirement can exploit some interference to enhance the legitimate user's secrecy rate.

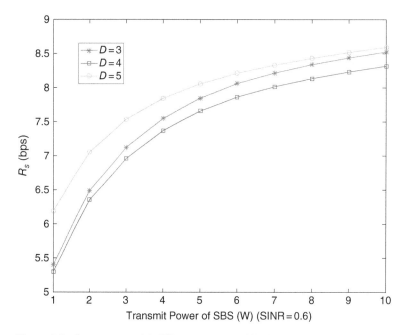

Figure 4.4 Secrecy rate with different number of D2D pairs.

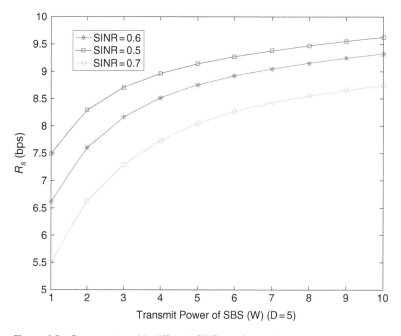

Figure 4.5 Secrecy rate with different SINR requirements.

4.5 Conclusion

This chapter introduced an effective security solution for 5G wireless networks using the physical layer security approach. The proposed solution relies on interference management techniques. A security analysis was conducted for the proposed interference management mechanism for the downlink in a two-tier HetNet, which includes beamforming, spectrum reuse, and D2D communications. In contrast to traditional interference management mechanisms that eliminate all interference, the proposed interference management mechanism can leverage beamforming and selective interference from spectrum reuse and D2D communications to enhance the security of communications in HetNet. Interference from SBSs to the user association with MBS can be strategically used to degrade the SINR of the eavesdropper while ensuring that the SINR requirement of other legitimate users is met. Additionally, D2D communications applied in the proximity of the eavesdropper can further reduce their SINR. However, it is important to note that not all interference in the network is beneficial for improving the secrecy rate.

5

Flexible and Efficient Security Schemes for IoT Applications in 5G Wireless Systems

Recent advancements in communication technologies and integrated circuits have fueled the widespread adoption and integration of Internet-of-Things (IoT) applications, creating a pervasive environment rich in both challenges and opportunities. This chapter introduces a comprehensive system architecture for IoT applications leveraging 5G wireless networks. The architecture considers the heterogeneity of IoT devices, users, network operators, and service providers. Various trust models are proposed and evaluated based on the trust relationships among these entities. Building upon the previously discussed system architecture and trust model, a secure and flexible authentication and data transmission scheme is proposed for heterogeneous IoT devices using 5G wireless systems [Fang et al., 2020].

5.1 IoT Application Models and Current Security Challenges

Ensuring security and privacy for IoT applications is a widely researched topic. However, given the diverse range of IoT application scenarios, creating a comprehensive system model for general IoT applications is challenging. Additionally, trust relationships between entities can vary depending on the specific application. In Kumar et al. [2017], an anonymous and secure framework for connected smart homes was proposed. The system model used was similar to that of wireless sensor networks, with the exception that the gateway is connected to the internet. However, previous studies have not adequately addressed the heterogeneity of IoT devices. Additionally, various scenarios for IoT device authentication have not been thoroughly investigated. In Song et al. [2017], only IoT devices and a server were considered in the system model for securing communications between these two parties. In Zhang et al. [2017a], the authors proposed a mobile-health system model with three entities as network manager, wireless body area network clients, and medical service providers. Trust relationships between these entities were discussed in the proposed protocol. However, the heterogeneity of IoT devices is not considered in the system.

In addition to the IoT system model, IoT devices are vulnerable to various types of attacks such as physical, side-channel, and cloning attacks due to their physical exposure and easy accessibility. Physical unclonable functions (PUFs) are considered to be built into IoT devices to provide security against those attacks. A PUF is a digital fingerprint that serves as a unique identity for a semiconductor device, which can derive a secret from the physical characteristics of the integrated circuits. Two primary applications of PUFs are

5G Wireless Network Security and Privacy, First Edition. Dongfeng (Phoenix) Fang, Yi Qian, and Rose Qingyang Hu.
© 2024 John Wiley & Sons Ltd. Published 2024 by John Wiley & Sons Ltd.

strong authentication and cryptographic key generation [Herder et al. 2014]. Considerable research has been devoted to assessing the properties of PUF hardware designs [Maiti et al., 2012, Yan et al., 2017b, Gao et al., 2017, Cao et al., 2018b]. Furthermore, there have been efforts to develop authentication protocols using PUFs. For instance, [Aman et al., 2017] proposed a mutual authentication and key exchange protocol based on PUFs. The protocol was designed to operate in two scenarios: between an IoT device and a server, and between two IoT devices. The proposed mutual authentication protocol relied on strong PUFs to achieve secure authentication. The protocol generated a session key using two random numbers, each generated by one of the participating entities. The research demonstrated that the protocol effectively guarded against cloning and physical attacks by utilizing PUFs. However, it should be noted that the protocol only considered IoT devices and servers.

Given the constraints of computational power and storage capacity of IoT devices, as well as privacy concerns, the current focus of research in IoT applications is on developing lightweight, efficient, and anonymous authentication methods. In Gope et al. [2018], the authors proposed a lightweight and privacy-preserving authentication protocol for a radio frequency identification (RFID) system. The protocol was designed to operate in an ideal PUF environment and protect against various types of attacks, including DoS, eavesdropping, impersonation, and cloning attacks. However, these PUF-based authentication methods are limited in their applicability to IoT systems as they only consider authentication between a tag and a server, and do not generate security keys post-authentication. Several studies, such as Liu et al. [2014], Xiong [2014], Zhao [2014], and He et al. [2017], have proposed anonymous authentications. Although in [Saeed et al., December 2018] the authors proposed a cost-effective and scalable authentication protocol, the previous works did not consider the comprehensive trust model and the heterogeneity of IoT devices. Therefore, this chapter aims to explore flexible and efficient security schemes for IoT applications that consider the comprehensive trust model and account for the heterogeneity of IoT devices.

5.2 A General System Model for IoT Applications Over 5G

This section presents a general IoT system architecture along with the corresponding trust models, threat models, and design objectives.

5.2.1 System Architecture

The architecture of the IoT system comprises four main components: heterogeneous IoT devices (classified as T1 and T2 IoT devices), IoT users, a network operator, and an IoT service provider. This is illustrated in Figure 5.1, where the two types of heterogeneous IoT devices are differentiated based on their distinct capabilities.

It is assumed that T1 IoT devices can directly connect to the network operator using either 3GPP or non-3GPP access technologies, which allows them to communicate directly with the provider. T1 IoT devices offer superior communication and computational capabilities compared to T2 IoT devices, and come equipped with a built-in security scheme. In contrast, T2 IoT devices are unable to establish a direct connection with the network operator.

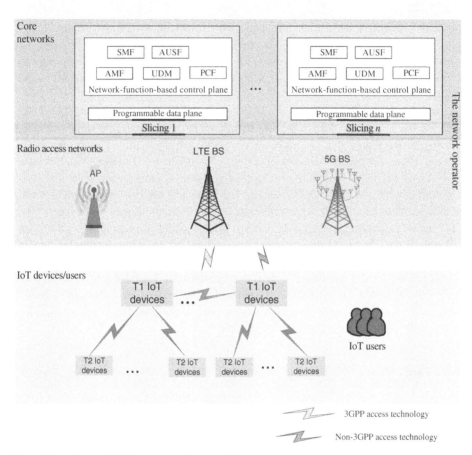

The IoT service provider

Data storage/analysis

Figure 5.1 A general IoT system model over 5G networks.

In addition, it is assumed that T2 IoT devices do not communicate with each other for the purposes of this study. Due to their limited battery life and computational capabilities, T2 IoT devices are unable to support complex security protocols. As such, a T1 IoT device is required to serve as a relay node for T2 IoT devices to connect to the network operator using non-3GPP access technologies.

Secure communication channels must be established between all entities involved. As depicted in Figure 5.1, the network operator is responsible for managing both the radio access networks and core networks. In 5G wireless systems, network slicing can be utilized to manage the resources based on specific applications dynamically. 5G core network differs from their 4G counterparts by segregating the control plane and data plane, which allows for improved performance and greater network flexibility during deployment [Fang et al., 2017a, 2017b, 2018]. In the control plane, different network functions are allocated, such as the access and mobility management function (AMF), session management function

(SMF), unified data management (UDM), authentication server function (AUSF), and policy control function (PCF) [Fang et al., 2017a]. The programmable data plane makes core network more flexible and better support different use cases in 5G.

The IoT service provider provides data storage and possible data analysis services for IoT users. The IoT service provider can also manage the identities of IoT devices if needed. IoT users can communicate with IoT devices and the corresponding IoT service provider. The communications between IoT users and IoT devices can be wireless-based or through the network operator. IoT users can communicate with the IoT service provider for data access and analysis services.

In the system model, there are assumed to be I T1 IoT devices, J T2 IoT devices, N IoT users, one IoT service provider, and one network operator. Typically, the value of J is significantly greater than I. A T1 IoT device D_i^1 can communicate with IoT users and the IoT service provider through the network operator. A T2 IoT device D_j^2 needs to go through a T1 IoT device to upload data to the IoT service provider and to communicate with other T1 IoT devices and IoT users. T1 IoT devices, IoT users, and the IoT service provider must first register with the network service provider before the communications between them. T2 IoT devices must first register with a T1 IoT device to join the system. Mutual authentication and initial session key agreement are required between two entities before they communicate with each other. T2 IoT devices must achieve mutual authentication with the IoT service provider to be able to connect to the network. Through a T1 IoT device, T2 IoT devices can join the network after the mutual authentication between T2 IoT devices and the IoT service provider. Moreover, all data transmission should be protected with data confidentiality, data integrity, and contextual privacy.

5.2.2 Trust Models

Trust models may vary depending on the specific applications in 5G wireless systems, which are used for a wide range of purposes. This study considers three potential trust models based on the IoT system model architecture presented in Section 5.2.1, which takes into account the varying degrees of trust among the different entities discussed in the previous section. Given that T1 IoT devices, IoT users, and the IoT service provider communicate directly with the network operator, these entities can be collectively regarded as "network entities." Figure 5.2 illustrates a trust model for 5G networks based on the IoT system model.

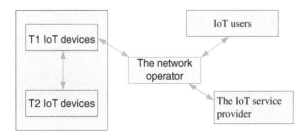

Figure 5.2 A trust model for 5G networks based on the IoT system model.

T2 IoT devices rely on T1 IoT devices to relay messages to other entities, but to maintain end-to-end security, T1 IoT devices must not be able to interpret messages exchanged between T2 IoT devices and other network entities. Various scenarios can be categorized according to the trust relationships among IoT users, T1 IoT devices, the network operator, and the IoT service provider, as outlined below.

Case 1: In this scenario, the network operator is deemed completely trustworthy by all other entities. Similarly, IoT users and IoT devices fully trust the IoT service provider. As such, the system model can treat the IoT service provider and network operator as a single entity. This configuration, known as Case 1, is commonly employed in early-stage research.

Case 2: Both the network operator and the IoT service provider are semi-trusted by IoT devices and IoT users. The network operator is fully trusted by the IoT service provider. The network operator and the IoT service provider are honest but curious about the data transmitted between IoT devices and IoT users. In this case, the IoT service provider can store information of IoT devices in the core network AUSF to achieve better latency management and mobility support. The system model can consider the IoT service provider and the network operator as one entity. Case 2 is also popularly used for simplifying the system model.

Case 3: Both the network operator and the IoT service provider are semi-trusted by IoT devices and IoT users. The network operator and the IoT service provider are honest but curious about the data transmitted between IoT devices and IoT users. The IoT service provider only uses the network operator services but not trusts it to manage information of IoT devices. In this case, the IoT service provider can deploy some servers close to the IoT devices and users to shorten the communication latency. This case is not well studied but is more practical to real applications.

5.2.3 Threat Models and Design Objectives

Out of the available trust models, Case 3 model has been selected for this study due to its minimal trust requirements between entities. In light of the fact that T2 IoT devices are susceptible to various forms of attacks such as physical, side-channel, and cloning attacks, it is assumed that these devices have weak PUFs. As a result, the IoT service provider only needs to store a single pair (C, R) for each T2 device. We assume that eavesdropping attacks can be launched to get communication information between IoT devices and other network entities. An identity stolen attack can be launched by cloning the identities of IoT devices to connect to the network. Also consider that the network operator can capture authentication messages and service data of IoT devices, which can be used to reveal sensitive information about IoT users. The IoT service provider, offering various services to IoT devices and IoT users, has strong incentives to protect the service data to attract more IoT users to use their services. On the other hand, the IoT service provider is curious about the personal information of users for advertisement recommendations and sensitive information IoT users share with their collaborators. Consequently, it is assumed that the IoT service provider has the capability to disclose users' personal information to third-party entities.

Building upon the Case 3 trust model and the threat model outlined earlier, the objective of this study is to develop a secure and adaptable authentication and data transmission

Table 5.1 Design objectives.

Security services	Features
Mutual authentication	To ensure that each entity is who they claim to be, it is imperative that any two communicating entities within the system authenticate each other
Initial session key agreement	After successful authentication, it is essential that an initial session key is only shared between the two verified entities
Data confidentiality and integrity	When transmitting data between two entities, it is critical that the confidentiality of the data is maintained, ensuring that the privacy of the data is not compromised, and that the data remains unaltered throughout the transmission
Anonymity	It is ensured that the intended recipient is the only party that can access the sender's identity
Contextual privacy	It is ensured that any adversary is unable to link the source and destination of a message
Forward security	Even if an entity's private key is exposed, it will not compromise the transmission that was safeguarded by the preceding session key
End-to-end security	Only the two communicating entities can access the transmitted data, ensuring that it remains unreadable to any other party
Key escrow resilience	Any attempt by the network operator to impersonate other entities will be detected, preventing them from doing so

scheme that ensures the safeguarding of privacy and security. The proposed scheme's design objectives are detailed in Table 5.1.

5.3 The 5G Authentication and Secure Data Transmission Scheme

This section presents an investigation into an flexible and efficient scheme for authentication and secure data transmission, aiming to achieve mutual authentication and the agreement of an initial session key while maintaining anonymity. In addition to authentication, this scheme incorporates data confidentiality, data integrity protection, and contextual privacy in the transmitted data. The scheme also ensures forward security, end-to-end security, and key escrow resilience. Before delving into the specifics of the proposed scheme, an overview will be provided.

5.3.1 Overview of the 5G Authentication and Secure Data Transmission Scheme

To achieve the design goals while taking into account communication and computational efficiency, the proposed scheme offers distinct authentication methods for T2 IoT devices and other entities.

The system initialization is the first step, which requires T1 IoT devices, IoT users, and the IoT service provider to register with the network operator to generate their respective private and public keys. The second step is the authentication phase, which requires mutual authentication between the two communicating entities before any data transmission can take place. After successful authentication, the generation of an initial session key is necessary. The authentication of non-T2 IoT devices differs from T2 IoT devices authentications. Before T2 IoT devices can communicate with any other entities, mutual authentication with a T1 IoT device and the IoT service provider is required. Mutual authentication between T2 IoT devices and the IoT service provider can be facilitated through the use of a T1 IoT device as a relay node. Depending on the application, the IoT service provider can simultaneously authenticate multiple T2 IoT devices. Once T2 IoT devices have joined the network, the relay T1 IoT device can assist in achieving mutual authentication between them and other entities. However, it is essential to establish an initial session key without disclosing it to the relay T1 IoT device.

Once authentication is complete, data transmission between these entities is facilitated with due consideration to data confidentiality, integrity protection, and contextual privacy. The proposed scheme also provides forward security, end-to-end security, and key escrow resilience, based on the trust model.

5.3.2 The Detailed Scheme

The flexible and efficient authentication and secure data transmission scheme includes five phases, system initialization, authentication and initial session key agreement, data transmission, data receiving, and T2 IoT devices authentication and initial session key agreement.

5.3.2.1 Phase 1 – System Initialization

Phase 1 involves the initialization of the system, which comprises two primary steps – system parameter generation and registration.

a) System parameter generation: The network operator generates two large prime numbers, p and q, based on a security parameter k that determines the input size of the computational problem. Let P be a generator of a cyclic group G with order q on elliptic-curve cryptography (ECC) [Stallings, 2017]. The network operator randomly generates $x_n \in Z_q^*$ as its own master private key and computes the public key $X_N = x_n P$. Additionally, the network operator chooses six secure hash functions: $H_0 : \{0,1\}^* \times Z_q^* \to Z_q^*$, $H_1 : \{0,1\}^* \times G \times G \times G \to Z_q^*$, $H_2 : \{0,1\}^* \times G \times G \times G \times \{0,1\}^* \to Z_q^*$, $H_3 : G \to \{0,1\}^* \times G \times G \times \{0,1\}^* \times \{0,1\}^*$, $H_4 : G \times \{0,1\}^* \times \{0,1\}^* \to Z_q^*$, $H_5 : G \times G \times Z_q^* \to \{0,1\}^*$. The network operator publishes the system parameter as $params = (p, q, P, X_N, H_0, H_1, H_2, H_3, H_4, H_5)$.
b) Registration: The registrations process for joining the system involves two distinct stages. During the first stage, T1 IoT devices, IoT users, and the IoT service provider register directly with the network operator. In the second stage, T2 IoT devices register with their corresponding relay T1 IoT device to gain access to the system.

For the network operator, T1 IoT devices (D^1), IoT users (U), and the IoT service provider (S) are network entities. Let W be a network entity, where $W \in \{D^1, U, S\}$. The registration steps for W are as follows.

- W randomly selects $x_w \in Z_q^*$ as a partial secret value and computes $X_w = x_w P$ as its corresponding partial public key.
- W sends its identity and the partial public key (W, X_w) to the network operator for registration.
- The network operator randomly chooses $y_w \in Z_q^*$ and computes $Y_w = y_w P$ and $z_w = y_w + x_n H_1(W||X_w||Y_w||X_n)$ to register W with the partial public key X_w.
- The network operator sends the partial private key z_w and partial public key Y_w through secure channel to W and stores the public key (X_w, Y_w).

Once W receives (z_w, Y_w), it can verify the validity of these two keys by comparing $z_w P$ with $Y_w + X_n H_1(W||X_w||Y_w||X_n)$. Then W stores the full private key (x_w, z_w) and the public key (X_w, Y_w).

For T2 IoT devices to join the system, they must register at a relay T1 IoT device. Assume the identity of ith T2 IoT device is D_i^2. R represents the relay T1 IoT device. Then the relay T1 IoT device computes $l_i = H_0(D_i^2||x_r)$ and $PD_i^2 = H_5(X_r||Y_r||l_i)$. Both l_i and PD_i^2 will be programmed into the T2 IoT device before deploying it into the network.

5.3.2.2   Phase 2 – Authentication and Initial Session Key Agreement

To establish a secure communication channel between any two of the network entities W_1 and W_2, mutual authentication is required, followed by the creation of a session key. The steps involved in this process are as follows.

- W_1 chooses a random number $a \in Z_q^*$, and picks up current time ft and computes $h_1 = H_1(W_2||X_{w_2}||Y_{w_2}||X_n)$, $n_1 = aP$, $h_2 = H_2(W_1||X_{w_1}||Y_{w_1}||n_1||ft)$, $Q_1 = h_2 P$, $Q_2 = H_3(h_2(X_{w_2} + Y_{w_2} + X_n h_1)) \oplus (W_1||X_{w_1}||Y_{w_1}||n_1||ft)$. W_1 sends the message (Q_1, Q_2) to the W_2.
- W_2 computes $h_3 = H_3((x_{w_2} + z_{w_2})Q_1)$ and gets $W_1||X_{w_1}||Y_{w_1}||n_1||ft = h_3 \oplus Q_2$. Then W_2 verifies the freshness ft and $h_2 P = Q_1$, chooses a random number $b \in Z_q^*$ and picks up current time $(ft)'$, and computes $n_2 = bP$, $h_1' = H_1(W_1||X_{w_1}||Y_{w_1}||X_n)$ $h_2' = H_2(W_2||X_{w_2}||Y_{w_2}||n_2||(ft'))$, $Q_1' = h_2' P$, $Q_2' = H_3(h_2'(X_{w_1} + Y_{w_1} + X_n h_1')) \oplus (W_2||X_{w_2}||Y_{w_2}||n_2||(ft)')$. W_2 sends the message (Q_1', Q_2') to W_1 and generates the initial session key $k_0^{w_1,w_2} = abP$.
- Once W_1 receives the reply (Q_1', Q_2') from W_2, W_1 performs the following steps: W_1 computes $h_3' = H_3((x_{w_1} + z_{w_1})Q_1')$ and gets $W_2||X_{w_2}||Y_{w_2}||n_2||(ft)' = h_3' \oplus Q_2'$. After verifying the freshness $(ft)'$ and $h_2' P = Q_1'$, W_1 generates the initial session key $k_0^{w_1,w_2} = abP$.

The authentication between T2 IoT devices and the IoT service provider is discussed in phase 5.

5.3.2.3   Phase 3 – Data Transmission

Data transmission in the system involves two types: one is between network entities, and the other is between T2 IoT devices and a single network entity.

a) Data transmission between network entities: To transmit a message m from network entity W_1 to network entity W_2, the message must be routed through the network operator. Both network entities will first find the session key of the current session t as $k_t^{w_1,w_2}$. Subsequently, the sender entity, W_1, will undertake the following steps.

- The sender W_1 randomly chooses $g \in Z_q^*$ and computes $h_1 = H_1(W_2||X_{w_2}||Y_{w_2}||X_n)$, $v_1 = gX_{w_2}$;
- To ensure the data source identity and integrity, W_1 computes $s_1 = gP, s_2 = H_4(s_1||W_1||m) + H_4(v_1||W_2||m), s_3 = g/(x_{w_1} + z_{w_1} + s_2)$;
- To ensure the data confidentiality, W_1 computes $v_2 = g(Y_{w_2} + h_1X_n), m' = H_5(v_1||v_2||k_t^{w_1,w_2}) \oplus m$;
- The ciphertext is (s_1, s_2, s_3, m', t).

W_1 sends the data as $(E_{X_{w_2}}(W_1), (s_1, s_2, s_3, m', t))$ through the network operator, and refreshes the session key between W_2 as $k_{t+1}^{w_1,w_2} = H_0(t||k_t^{w_1,w_2})$.

b) Data transmission between a T2 IoT device and other network entities is discussed in detail in phase 5 after the T2 IoT devices join in the network.

5.3.2.4 Phase 4 – Data Receiving

When W_2 receives $(E_{X_{w_2}}(W_1), (s_1, s_2, s_3, m', t))$, W_2 decrypts the identity $D_{X_{w_2}}(E_{X_{w_2}}(W_1))$ and finds the session key with W_1 $k_t^{w_1,w_2}$ and the public key of W_1 (X_{w_1}, Y_{w_1}). Then the following steps are executed to decrypt and verify the received ciphertext.

- W_2 computes $v_1' = x_{w_2}s_1, v_2' = z_{w_2}s_1, m'' = H_5(v_1'||v_2'||k_t^{w_1,w_2}) \oplus m'$
- Computes $H_4(s_3(X_{w_1} + Y_{w_1} + s_2P)||W_1||m'') + H_4(v_1'||W_2||m'')$. If $H_4(s_3(X_{w_1} + Y_{w_1} + s_2P)||W_1||m'') + H_4(v_1'||W_2||m'') = s_2$, the message m'' is accepted.

W_2 refreshes the session key between W_1 as $k_{t+1}^{w_1,w_2} = H_0(t||k_t^{w_1,w_2})$.

5.3.2.5 Phase 5 – T2 IoT Devices Authentication and Initial Session Key Agreement

T2 IoT devices must pass three levels of authentication. Firstly, they must establish mutual authentication with the relay T1 IoT device to enable a connection after registration. Secondly, through the T1 IoT device, T2 IoT devices must achieve mutual authentication with the IoT service provider to be granted system access. This authentication uses a PUF function to verify the legitimacy of both the T2 IoT device and the IoT service provider. Thirdly, once connected to the system, T2 IoT devices must undergo mutual authentication with other network entities. The relay T1 IoT device facilitates the second and third levels of authentication.

For the authentication between a T2 IoT device and the relay T1 IoT device, the relay T1 IoT device sends its public key X_r to the T2 IoT device. The ith T2 IoT device randomly generates N_0 and sends $(PD_i^2, N_0, E_{X_r}(N_0||l_i))$ to the relay T1 IoT device. The relay T1 IoT device decrypts and verifies the message $H_0(N_0||l_i)$, and randomly generates N_1 and sends $(N_0 \oplus N_1, H_0(N_1||l_i),)$. The T2 IoT device finds out N_1 and verifies the $H_0(N_1||l_i)$. The T2 IoT device and the relay T1 IoT device establish an initial session key by computing $k_0^{D_i^2,r} = H_0((N_0 \oplus N_1)||l_i)$, which can be used to secure the communications between them.

After the mutual authentication between a T2 IoT device and the relay T1 IoT device, mutual authentication between the T2 IoT device and the IoT service provider is needed to allow the T2 IoT device to join the network. Through the relay node, the IoT service provider can authenticate the T2 IoT device through the following steps.

- The relay T1 IoT device sends message $m_1 = (D_1^2, D_2^2, \ldots, D_I^2)$ to the IoT service provider, where D_i^2 is the ith T2 IoT device connecting to the relay T1 IoT device.
- After the challenge message is sent by the T2 IoT device, the IoT service provider authenticates the message by verifying it with the corresponding device identity $C = ((C_1, D_1^2), (C_2, D_2^2), \ldots, (C_I, D_I^2))$. Upon successful verification, the IoT service provider sends a nonce N_0 along with the verified message to the relay T1 IoT device.
- The relay T1 IoT device sends each T2 IoT device the corresponding challenge $(C_i, E_{k_0^{D_i^2,r}}(N_0))$.
- Each T2 IoT device computes a hash value of $E_{k_0^{D_i^2,r}}(H_0(N_0||f(C_i)))$ and sends it to the relay T1 IoT device, where $f(C_i) \in Z_q^*$.
- The relay T1 IoT device decrypts the message, and sends $m_2 = (H_0(N_0||f(C_1)) \oplus H_0(N_0||f(C_2)) \oplus \cdots \oplus H_0(N_0||f(C_I)))$.
- The IoT service provider verifies the message, sends the confirmation to the relay T1 IoT device, and generates a secret key between itself and the ith T2 IoT device $k^{D_i^2,s} = H_0(D_i^2||H_0(N_0||f(C_i)))$.
- The relay T1 IoT device sends the confirmation to T2 IoT devices. Both the relay T1 IoT device and each T2 IoT device refresh their session key $H_0(N_0||k_0^{D_i^2,r})$.
- The ith T2 IoT device generates the session key between itself and the IoT service provider $k^{D_i^2,s} = H_0(D_i^2||H_0(N_0||f(C_i)))$.

After T2 IoT devices join the network, other network entities can communicate with T2 IoT devices through the relay T1 IoT device. The authentication process consists of the following steps.

- W initiates a communication request by sending a nonce N_0 and the pseudonym PD_i^2 of the T2 IoT device to the relay T1 IoT device.
- The relay T1 IoT device verifies the network entity and sends public key X_w and N_0 of the network entity W to the T2 IoT device.
- The T2 IoT device randomly generates N_1 and computes $E_{X_w}(N_1)$ and $H_0(l_i||E_{X_w}(N_1))$, which are sent to the relay T1 IoT device. The T2 IoT device generates a session key $k_0^{D_i^2,w} = H_0((N_0 \oplus N_1)||X_w)$.
- The relay T1 IoT device verifies $H_0(E_{X_w}(N_1)||l_i)$ and sends $E_{X_w}(N_1)$ to the network entity W.
- The network entity decrypts the message with its full private key and generates the session key $k_0^{D_i^2,w} = H_0((N_0 \oplus N_1)||X_w)$.

5.4 Security Analysis

This section provides an analysis on the security properties of the proposed scheme. Specifically, it evaluates the proposed protocol's ability to meet the design objectives in Section 5.2.

5.4.1 Protocol Verification

The first step is to verify the protocol for both the authentication and key setup procedures. Upon receiving the message (Q_1, Q_2) from W_1, W_2 first needs to decrypt the ciphertext using its full private key (x_{w_2}, z_{w_2}) as follows.

$$H_3((x_{w_2} + z_{w_2})Q_1) \oplus Q_2 = H_3((x_{w_2} + z_{w_2})h_2P) \oplus H_3(h_2(X_{w_2} + Y_{w_2} + h_1))$$
$$\oplus (W_1 || X_{w_1} || Y_{w_1} || n_1 || ft)$$
$$= H_3((X_{w_2} + Yw_2 + X_n h_1)h_2) \oplus H_3(h_2(X_{w_2} + Y_{w_2} + X_n h_1))$$
$$\oplus (W_1 || X_{w_1} || Y_{w_1} || n_1 || ft)$$
$$= (W_1 || X_{w_1} || Y_{w_1} || n_1 || ft),$$

where $h_1 = H_1(W_2 || X_{w_2} || Y_{w_2} || X_n)$ and $h_2 = H_2(W_1 || X_{w_1} || Y_{w_1} || n_1 || ft)$. Furthermore, the shared initial session key generation due to the fact that W_1 generates a and gets $n_2 = bP$ from W_2, and W_2 generates b and gets $n_1 = aP$ from W_1. Therefore, they can both generate the initial session key abP.

For the verification of data transmission protocol, once getting the message $(E_{X_{w_2}}(W_1)$, $(s_1, s_2, s_3, m', t))$, the receiver decrypts the identity of the sender with $D_{x_{w_2}}(E_{X_{w_2}}(W_1))$ first, then decrypts the ciphertext using its full private key (x_{w_2}, z_{w_2}) and current session key $k_t^{w_1, w_2}$ as follows.

$$m'' = H_5(v_1' || v_2' || k_t^{w_1, w_2}) \oplus m'$$
$$= H_5(x_{w_2} gP || z_{w_2} gP || k_t^{w_1, w_2}) \oplus H_5(gX_{w_2} || g(Y_{w_2} + h_1 X_n) || k_t^{w_1, w_2}) \oplus m$$
$$= H_5(gX_{x_2} || g(Y_{w_2} + h_1 X_n) || k_t^{w_1, w_2}) \oplus H_5(gX_{w_2} || g(Y_{w_2} + h_1 X_n) || k_t^{w_1, w_2}) \oplus m$$
$$= m.$$

Since $m'' = m$, the verification is as follows.

$$H_4(s_3(X_{w_1} + Y_{w_1} + s_2 P) || W_1 || m'') + H_4(v_1' || W_2 || m)$$
$$= H_4(g/(x_{w_1} + z_{x_1} + s_2)(X_{w_1} + Y_{w_1} + s_2 P) || W_1 || m) + H_4(x_{w_2} gP || W_2 || m)$$
$$= H_4(gP || W_1 || m) + H_4(gX_{w_2} || W_2 || m)$$
$$= s_2.$$

5.4.2 Security Objectives

This section verifies the achieved security objectives based on the proposed protocol.

5.4.2.1 Mutual Authentication

The proposed scheme can achieve mutual authentication for multiple scenarios, including situations where sender W_1 needs to authenticate itself to receiver W_2. To achieve this, W_1 generates a hash value using both its own private key and the full private key of W_2, which ensures that only W_2 can authenticate the message. Specifically, the proposed scheme ensures that only the receiver is able to recover the identity of the sender and authenticate itself to the sender using the same approach. The mutual authentication between a T2 IoT device and other entities is achieved through the help of a relay T1 IoT device. To join the system, the T2 IoT device must first undergo mutual authentication with the relay T1 IoT device. This is achieved using a pre-shared hash value that is known to both devices. Once authenticated, the T2 IoT device can proceed to authenticate itself to the IoT service provider. This is accomplished using a PUF function, with assistance from the relay T1 IoT device. Finally, to authenticate itself to other network entities, the T2 IoT device relies on the relay T1 IoT device.

5.4.2.2 Initial Session Key Agreement

After successful authentication, the two network entities W_1 and W_2 are able to share a secure session key. The key materials a and b are unique to the two network entities, meaning that they can only be shared between them. In other words, these key materials can only be shared by these two network entities. More specifically, $k_0^{w_1,w_2}$ can only be computed by the network entity W_2 with its full private key (x_{w_2}, z_{w_2}) and b which is generated by W_2. And $k_0^{w_1,w_2}$ can only be computed by the network entity W_1 with its full private key (x_{w_1}, z_{w_1}) and a which is generated by W_1. Using the key materials exchanged during the authentication phase, the proposed scheme can ensure that the two parties have correctly shared the session key.

5.4.2.3 Data Confidentiality and Integrity

The proposed scheme ensures data confidentiality and integrity through the use of several measures. Specifically, data confidentiality is protected using both a session key $k_t^{w_1,w_2}$ and the public key encryption. The sender W_1 encrypts the data using the public key of the receiver and the session key that is shared between the sender W_1 and the receiver W_2. The message m is encrypted to $m' = H_5(v_1||v_2||k_t^{w_1,w_2}) \oplus m$, where v_1, v_2 only can be recovered with the receiver's full private key (x_{w_2}, z_{w_2}). The session key $k_t^{w_1,w_2}$ is only shared and stored between the sender and the receiver.

Data integrity is guaranteed with the signature of the sender. By decrypting and verifying the ciphertext $(E_{X_{w_1}}(W_1), (s_1, s_2, s_3, m', t))$, with $H_4(s_3(X_{w_1} + Y_{w_1} + s_2P)||W_1||m'') + H_4(v_1'||W_2||m) = s_2$, the receiver can recover the message and check the integrity of the message.

5.4.2.4 Contextual Privacy

The proposed scheme achieves contextual privacy by ensuring that the source identity of a message is not transmitted in plaintext. Rather, only the intended receiver, using its private key, can retrieve the source identity of the message.

5.4.2.5 Forward Security

After the initial session key generation, a new session key is generated at the end of each transmission, which helps to ensure forward secrecy. To generate the new session key, a hash function is applied to the current session key. Due to the one-way property of the hash function, attackers cannot recover the previous session key even if they get the current session key. Moreover, even if the full private key of the network entity is revealed, the previous transmission remains confidential since the previous session key is not able to be recovered.

5.4.2.6 End-to-End Security

The proposed scheme ensures end-to-end security, meaning that messages exchanged between any two entities are not accessible to the network operator as they lack a partial private key generated by the user. The full private key and the session key are required to decrypt messages, and only the receiver possesses both. Additionally, the session key is exclusively known by the communicating entities, ensuring that end-to-end security is maintained.

5.4.2.7 Key Escrow Resilience

The proposed scheme features a private key generation process that involves both the network operator and the network entity. Each entity's private key consists of a partial key generated by the entity itself, and another partial key generated by the network operator. This dual-generation approach ensures that the network operator cannot impersonate any other entity undetected, as the entity-generated partial key acts as a safeguard against such malicious attempts.

5.5 Performance Evaluation

This section evaluates the performance of the proposed scheme in three key areas: security services provided, computational overhead, and communication overhead. By analyzing these metrics, the effectiveness of the scheme can be accessed in providing robust security while minimizing the computational and communication costs associated with its implementation.

5.5.1 Security Services

This section compares the security services provided by the proposed scheme with those of other similar schemes. It's worth noting that most existing authentication schemes are limited to a single scenario involving one type of IoT device and a server, due to their simplified system models. However, the proposed scheme overcomes this limitation by providing a comprehensive set of security services that are adaptable to a wide range of scenarios and devices within the IoT ecosystem. The proposed mutual authentication scheme is highly versatile and can be adapted to suit a variety of scenarios, thanks to the proposed system model architecture. Additionally, the scheme goes beyond simple authentication to provide a range of robust security services, including confidentiality and integrity, and contextual privacy of data transmission. These measures help ensure the privacy of sensitive information during transmission. Further details of the security services comparison can be found in Table 5.2, where 1 = [Saeed et al., 2018], 2 = [Liu et al., 2014], 3 = [Xiong, 2014], 4 = [Zhao, 2014], 5 = [Wang and Zhang, 2015], and 6 = [He et al., 2017].

5.5.2 Computational Overhead

Table 5.3 provides an efficiency comparison between the proposed scheme and the existing authentication schemes proposed in Saeed et al. [2018], Liu et al. [2014], Xiong [2014],

Table 5.2 Security services comparison.

Schemes	Mutual authentication	Contextual privacy	Data confidentiality	Forward security	End-to-end security	Key escrow resilience
1	Yes (single scenario)	No	Yes	Yes	Yes	Yes
2	Yes (single scenario)	No	Yes	No	Yes	Yes
3	Yes (single scenario)	No	Yes	Yes	Yes	Yes
4	Yes (single scenario)	No	Yes	No	Yes	Yes
5	Yes (single scenario)	No	Yes	Yes	Yes	Yes
6	Yes (single scenario)	No	Yes	Yes	Yes	Yes
The proposed	Yes (multiple scenario)	Yes	Yes	Yes	Yes	Yes

Zhao [2014], Wang and Zhang [2015], and He et al. [2017]. This comparison is based on two key metrics: computational overhead and computational time cost. An analysis of these metrics provides valuable insights into how the proposed scheme compares to existing schemes in terms of efficiency. This, in turn, enables to assess the effectiveness of the proposed scheme in delivering secure, efficient authentication services within the context of the IoT ecosystem. Since the overhead of arithmetic operations and hash functions can be ignored compared to paring operation, exponentiation, and point multiplication, only paring operation, exponentiation, and point multiplication are considered. In Table 5.3, p, e, $*$ represent a pairing computation in G, an exponentiation in G, and a point multiplication in G.

According to the experimental results in Aranha et al. [2010], 0.32 s on ATmega 128L-based platform is needed to achieve an 80-bit security level on the ECC-based scalar multiplication on curve $E : y^2 = x^3 + ax^2 + b$. To achieve the same security, pairing takes 1.90 s, and point multiplication needs 0.81 s on ATmega 128L-based platform [Oliveira et al., 2011]. Table 5.3 presents a detailed comparison of the computational overhead and computational time cost associated with the proposed scheme and the existing authentication schemes. These metrics provide key insights into the efficiency of the proposed scheme and its ability to deliver secure, efficient authentication services within the IoT

Table 5.3 Computational overhead and time cost comparison.

Schemes	Computational overhead	Computational time cost	Curve type require
1	$6*+e$	5.76 s	Pairing-based
2	$6*+e$	4.14 s	Pairing-based
3	$10*$	3.20 s	ECC-based
4	$e+4*$	2.18 s	ECC-based
5	$3*+p$	4.33 s	Pairing-based
6	$4*$	3.24 s	Pairing-based
The proposed	$6*$(for others), $2e$(for T2 IoT devices)	1.92 s, 1.8 s	ECC-based

ecosystem. The proposed authentication scheme requires less computational overhead and time cost for network entities and T2 IoT devices. Since the authentication of T2 devices can be processed in a batch fashion, the IoT service provider can authenticate T2 IoT devices more efficiently compared to other schemes. The computational overhead comparison between the proposed scheme and the other schemes is shown in Figure 5.3 with 50% T2 IoT devices and Figure 5.4 with 70% T2 IoT devices. From Figures 5.3 and 5.4,

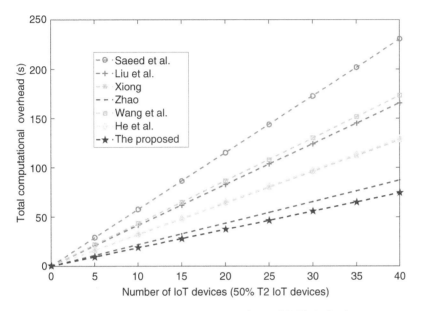

Figure 5.3 The computational overhead comparison (50% T2 devices).

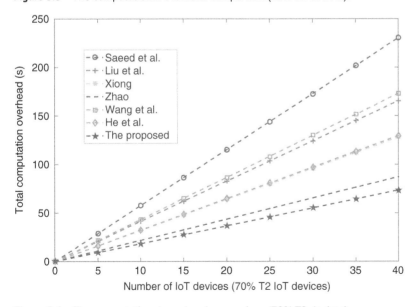

Figure 5.4 The computational overhead comparison (70% T2 devices).

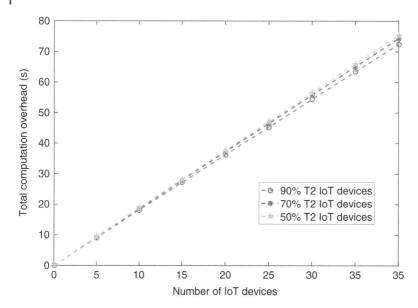

Figure 5.5 The computational overhead comparison of the proposed scheme based on different percentages of T2 IoT devices.

the computational overhead of the proposed scheme is less than the existing schemes. The computational overhead comparison based on a different percentage of T2 IoT devices is shown in Figure 5.5, which shows that with the same total number of IoT devices, the more percentage of T2 IoT devices, the less computational overhead will be needed. However, due to the small difference in the computational time cost between T1 and T2 IoT devices, the computational overhead of the proposed scheme does not have a significant difference with different percentages of T2 IoT devices, as shown in Figure 5.5.

5.5.3 Communication Overhead

The communication overhead is compared with the existing authentication schemes as shown in Table 5.4 in terms of transmit overhead and receive overhead.

Based on the calculation in Saeed et al. [2018], the curve used in pairing with compressed G_1 size is 34 bytes, and G_2 size is 136 bytes. For ECC-based, the compressed G size is 21 bytes. A comparison of the communication overhead of difference schemes in terms of these parameters is shown in Table 5.4. The proposed scheme has significantly low overhead for T2 IoT devices. The communication overhead comparison is shown in Figure 5.6 with 50% T2 IoT devices and in Figure 5.7 with 70% T2 IoT devices. Figures 5.6 and 5.7 show that the proposed scheme has the least communication overhead compared to other similar schemes. The communication cost comparison of the proposed scheme based on different percentages of T2 IoT devices is shown in Figure 5.8, which indicates that with the same total number of IoT devices, the higher percentage of T2 IoT devices, the less communication cost will be needed.

Table 5.4 Communication overhead comparison.

Schemes	Transmit overhead	Receive overhead
1	$3|Z_p^*|| +4|G_1| + |right| + |t|$, 242 bytes	$|MAC| + |G_1| + t$, 68
2	$|Z_p^*| + 2|G_1| + 2|G_2| + |right| + |t|$, 382 bytes	$|MAC|$, 32
3	$|ID| + 4|G| + |t|$, 90 bytes	$|MAC| + |G|$, 53
4	$3|G| + |ID| + |right| + |t|$, 77 bytes	$|G| + |MAC|$, 53
5	$|ID| + 2|G_1| + 2|t|$, 76 bytes	$|MAC| + |G_1| + t$, 68
6	$|ID| + 2|G_1| + |right| + |t|$, 82 bytes	$|MAC| + |G_1|$, 66
The proposed	$|W| + 4|G| + |t|$, 90 bytes (for others) $|Z_p^*|$, 32 bytes (for T2 IoT devices)	$|W| + 4|G| + |t|$, 90 bytes $|Z_p^*|$, 32 bytes

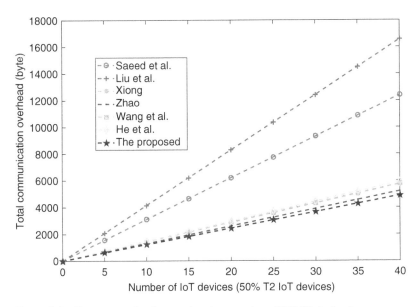

Figure 5.6 The communication overhead comparison (50% T2 devices).

5.6 Conclusion

This chapter begins by examining a general IoT system architecture over 5G, using three distinct trust models. The studied system model takes into account the heterogeneity of IoT devices, enabling to design a flexible and efficient authentication

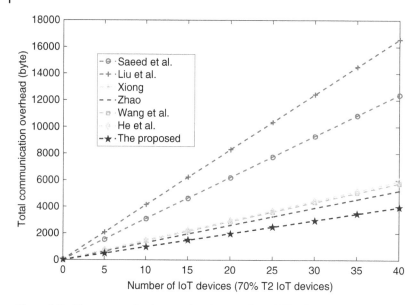

Figure 5.7   The communication overhead comparison (70% T2 devices).

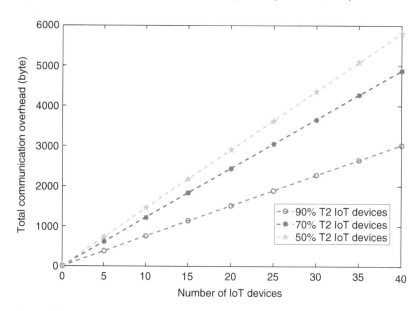

Figure 5.8   The communication overhead comparison of the proposed scheme based on different percentages of T2 IoT devices.

and secure data transmission scheme based on the third trust model. The proposed authentication and secure data transmission scheme provides anonymous mutual authentication for multiple scenarios, while also ensuring data transmission confidentiality, integrity, and contextual privacy, in order to meet the needs of new use cases over 5G wireless systems. Session key agreement, forward security, end-to-end security, and key escrow resilience are also achieved with the proposed scheme. A security analysis is presented to verify the functionality of the proposed scheme. Performance evaluations in terms of security services, computational overhead, and communication overhead are illustrated to show that the proposed scheme achieves security services and offers less overhead to T2 IoT devices with flexibility and efficiency.

6

Secure and Efficient Mobility Management in 5G Wireless Networks

This chapter introduces a novel mobility management scheme for 5G wireless networks based on software-defined networking (SDN) architecture. The proposed network architecture simplifies mobility management and enhances efficiency. Additionally, the implementation of dual connectivity technology enables fast authentication by separating control plane and user plane handover. The chapter delves into three different handover scenarios and their respective procedures. Furthermore, it proposes two authentication protocols, a full authentication protocol and a fast authentication protocol, for user registration and handover scenarios.

6.1 Handover Issues and Requirements Over 5G Wireless Networks

Numerous research activities have been conducted to enhance handover performance, primarily by minimizing the frequency of handovers [Shen and van der Schaar, 2017, Chen et al., 2017, Semiari et al., 2017, Tang et al., 2016]. According to Lauridsen et al. [2017], in long-term evolution (LTE) networks, each handover implementation results in a median-level data interruption of 40 ms. It is evident that reducing the handover frequency alone cannot meet the service requirements of specific 5G applications as handover latency remains a concern. In cases where handover frequency is dictated by service requirements, reducing the frequency is not a viable solution. Thus, there is a need to focus on reducing the latency associated with each handover.

Handover latency can be divided into control plane (CP) latency and user plane (UP) latency. According to Lauridsen et al. [2017], the latency targets for CP and UP in LTE networks are 100 and 20 ms, respectively. The merger of CP and UP in LTE networks results in substantial signaling overhead during handover between RANs and the CN. Therefore, a decoupled CP and UP architecture is proposed in Prados-Garzon et al. [2016]. In Yan et al. [2017a], the authors explored the concept of decoupled CP and UP handover. However, the majority of the current literature is based on LTE core networks, without considering the 5G RAN and CN. It is crucial to reduce communication, computation overhead, and signaling overhead between RANs and the CN to enhance handover latency performance significantly.

5G Wireless Network Security and Privacy, First Edition. Dongfeng (Phoenix) Fang, Yi Qian, and Rose Qingyang Hu.
© 2024 John Wiley & Sons Ltd. Published 2024 by John Wiley & Sons Ltd.

The 5G RAN and CN demand greater flexibility, scalability, cost-effectiveness, and energy efficiency. In 5G CN, mobility management entity (MME) and home subscriber server (HSS) functions in LTE core networks will be replaced by SDN-based functions. Unlike in LTE networks, where CP and UP functions are executed in the serving gateway (S-GW) located far away from user equipment (UE), 5G CN architecture will enable more efficient execution of these functions. In 5G CN, access control and session management are separated to provide better support for fixed access, flexibility, and scalability. Unlike in LTE networks, where the CP and UP functions are performed in the S-GW, in 5G CN, the CP and UP separation in the S-GW reduces the cost of user plane, and brings the UP closer to the UE. This approach not only reduces the path switch time but also results in significant cost savings. The 5G RAN will employ new radios to enhance its performance. Additionally, the use of SDN can improve the base stations (BSs) by enabling certain functions to be executed within the CN. In Nayak et al. [2018], the authors proposed a centralized SDN architecture based on a service-based 5G CN. The architecture included new radio nodes aimed at reducing signaling between the RAN and CN. However, the new radio nodes are still similar to the eNB in LTE, and each handover still requires communication with the CN.

6.2 A 5G CN Model and HetNet System Model

This section presents the 5G CN model used in this study, including a comparison of its functions with those of LTE entities. Additionally, a HetNet system model, along with the corresponding trust and attack models, is introduced. To improve the robustness of the 5G system, dual connectivity is utilized.

Figure 6.1 presents a comparison between the LTE and 5G CNs. The left-hand side illustrates the LTE CN with its various entities, while the right-hand side depicts the

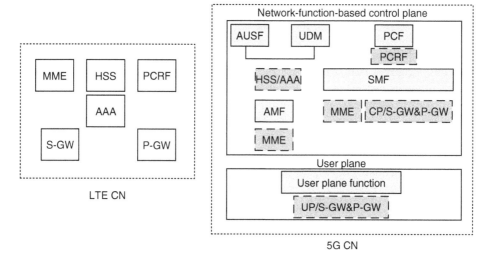

Figure 6.1 Comparison of an LTE CN and a 5G CN.

5G CN, which features a network function-based control plane and a user plane that is separated. The equivalent entities used in the 5G CN, as compared to the LTE CN, are also indicated in the figure. Compared to the LTE CN, the 5G CN separates the two functions in MME as access and mobility management function (AMF) and session management function (SMF). With these two functions separated, 5G wireless networks can better support fixed access and achieve higher flexibility and scalability. AMF is expected to be access-independent, which can simplify the mobility management between different radio access technologies (RATs). AMF is responsible for managing access control and mobility, and its application may vary depending on the use case. SMF can set up and manage sessions. Multiple SMFs can be assigned to manage different sessions of a single user associated with a single AMF. The control functions in the traditional S-GW and packet gateway (P-GW) are also moved in SMF in the 5G CN. Authentication server function (AUSF) and user data management (UDM) are implemented in the HSS and authentication, authorization, and accounting (AAA). The policy control function (PCF) achieves the same purpose as the policy and charging rule function (PCRF). All these network functions interact with each other with well-defined interfaces.

In LTE, depending on the RATs, MME or AAA, and HSS are involved in UE mobility management. Based on the 5G core network (CN), AMF, AUSF, and UDM are utilized to achieve mutual authentication and key agreement (AKA) between UE and core networks. In the handover between different cells, SMF can be involved in new session setup and management. The separation of the CP and UP in the 5G core network is expected to achieve flexible deployment and efficient management.

The proposed model for this study takes into account a HetNet system, which is illustrated in Figure 6.2. The system comprises two macro-BSs, each with several small access points (APs) that support different RATs such as 5G new radio and 3GPP radio. In LTE networks, macrobase stations are typically connected through wired connections due to their low density. However, with the introduction of 5G technology and higher-frequency bands, it will be possible to deploy macrobase stations at a higher density and establish wireless communication between them. 5G technology incorporates advanced features to support various use cases, including enhanced mobile broadband (eMBB), ultra-reliable low latency (URLLC), and massive machine-type communications (mMTC). Since these use cases have different requirements, such as low latency for URLLC, mobility may have varying impacts on their performance.

To improve the handover performance between macrocells, small-cell APs are deployed at the overlap area, as shown in Figure 6.2. The connections between BSs to APs are the *Xn* interface. In this system model, each AP must establish a secret sharing with the base stations BSs within its coverage area upon joining the network. To ensure both reliability and zero handover interruption time, the system employs dual connectivity. Each UE can have a connection with a small cell and a macrocell at the same time. The small cell can provide the UP connection for high data rate services. The macrocell connection provides a CP connection and also a backup UP connection. For the SDN-based core network, CP and UP are separated. The separation can significantly reduce the cost of the conventional gateway. Furthermore, deploying multiple UPs at the edge networks can bring them closer to the RAN. In the proposed network system architecture, each BS can handle handovers between APs within its coverage area. For inter-BS handovers, the APs located in the overlapping

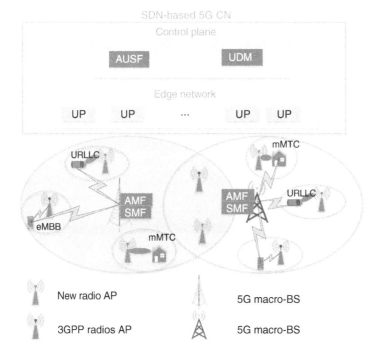

Figure 6.2 A 5G HetNet system model.

areas can assist in maintaining UP connections and facilitate seamless handover without interrupting regular services.

After analyzing the LTE CN and 5G CN, it has been identified that mobility management involves several parties, including the UE, AP, BS/AMF/SMF, AUSF, and UDM. Figure 6.3 illustrates that the BS, AMF, and SMF are treated as a single entity in this context. The long-term keys and real identities of UE are only known by UE and UDM. The communication links between UE and BS/AMF/SMF, between BS/AMF/SMF and AP, and between UE and AP are not secure, as shown in the dotted lines in Figure 6.3. The communication links between BS/AMF/SMF and AUSF, and between AUSF and UDM are secure, as shown in the solid lines in Figure 6.3. The BS/AMF/SMF, AP, and AUSF are semi-trust.

The studied system model considers various attack models as follows.

- Eavesdropping: As UE identities are transmitted in plaintext, an eavesdropping attack can reveal their actual identities, which are used for network registration, enabling an

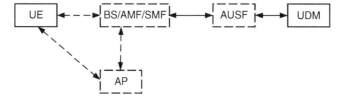

Figure 6.3 Trust model in the studied system model.

attacker to impersonate the UE. By utilizing the AKA protocol in LTE, the attacker can obtain location and even conversation information by forwarding the identities to the MME. Additionally, the attacker may launch man-in-the-middle (MITM) and DoS attacks.

- MITM: If the attacker acquires the UE actual identity, they can also obtain the random challenge (RAND) and authentication token (AUTN) from the MME. By forwarding these values to the UE, the attacker can obtain the appropriate response (RES). This would enable the attacker to impersonate a BS to the UE or a UE to a real BS.
- DoS: An attacker may repeatedly send bogus or different UE identities to the MME, which could exhaust the HSS computational resources when generating authentication vectors for the UE. Furthermore, the MME may be forced to reserve memory for an extended period while waiting for a response from the attacker.
- Replay attacks: By intercepting the communication between a UE and a BS, an attacker can repeatedly forward the messages to the BS, initiating a replay attack.

6.3 5G Handover Scenarios and Procedures

Considering the high density of small cells in 5G networks and their service-specific requirements, handover between higher frequencies is unavoidable. Therefore, ensuring secure and efficient mobility management is crucial to meet the 5G service requirements. Flexibility in small-cell deployment is essential to address different scenarios effectively. In scenarios such as public safety, additional small cells can be utilized to ensure high network throughput. Thus, the proposed model includes an analysis of handover mechanisms between small-cell APs and 5G macro-cell BSs. In order to enhance handover performance, AMF and SMF are deployed near the BSs, as depicted in Figure 6.2. This section examines various handover scenarios within the proposed 5G HetNet system model and presents the corresponding handover procedures based on these scenarios.

6.3.1 Handover Scenarios

Based on the illustrated 5G HetNet system model in Figure 6.2, each user can have dual connectivity with a macrocell BS, and a small-cell AP, where the BS connection can provide communication reliability and AP connection can ensure high data rate requirement. In each macrocell, UE can keep the BS connection with both CP and UP. Small-cell AP handover can be triggered due to signal quality or service requirements on a high data rate. When the UE gets closer to the macrocell BS, the UE may need to switch the UP from small-cell AP to macrocell BS due to the signal quality requirement. When the UE moves away from the macrocell BS, the UE may need to have the UP handover from the macrocell BS to a small-cell AP. When UE moves from one AP to another AP, the UP handover between these two APs can be triggered. When UE moves to the edge of the macrocell BS, a handover between the macrocell BS to another macrocell BS is required.

Therefore, the possible handover scenarios are:

- From a macrocell BS to a small-cell AP;

- Between two small-cell APs under the same macrocell BS;
- Between two 5G macrocell BSs.

In LTE networks, to achieve seamless handovers between the evolved universal terrestrial radio access network (E-UTRAN) and non-3GPP access networks, a full access authentication procedure between a UE and the target access network is required before the UE handover to the new access network, which can bring a significant handover delay through multiple rounds of message exchanges to an AAA server. The proposed system model in this study enables simplified inter-RAT handover, as the AMF are independent of the macrocell BSs and RATs. Additionally, the separation of the CP and UP handover, as well as the implementation of dual connectivity, can further enhance the handover performance.

6.3.2 Handover Procedures

The procedures for each of the three handover scenarios are presented below.

- Handover from a macrocell BS to a small-cell AP: The transition from a macrocell BS to a small-cell AP can occur when the UE moves out of range of the macrocell BS and obtains a stronger signal from a small-cell AP, or when the service demands a higher data rate. Even when the UE is within the coverage area of the macrocell BS, it maintains its connection with the macrocell BS, with both CP and UP as backup. Thus, only a UP handover is required when transitioning from the macrocell BS to a small-cell AP, as depicted in Figure 6.4.
The UE sends the measurement report and service requirement to the BS first. Based on the measurement report and service requirement, the BS can make the UP handover decision. The BS then sends a handover request to the target AP. Once the BS gets the handover request acknowledgment, the handover command will be sent to the UE. The UP handover procedures are managed within the RAN, without the need for involvement from the CN. During the UP handover, the CP connection between UE and the BS

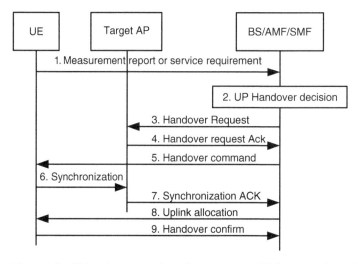

Figure 6.4 UP handover procedures from a macrocell BS to a small-cell AP.

has no change, which can ensure the reliability of the mobility management. Since the UP handover from the BS to the target AP is managed solely within the RAN, it incurs lower signaling overhead and has a shorter transmission path compared to LTE handovers involving the CN.

- Handover between two small-cell APs: Figure 6.5 illustrates that handovers between two small-cell APs operating under the same macrocell BS coverage may be frequently initiated due to the utilization of high-frequency radio spectrum, exceeding 10 GHz. Efforts to mitigate frequent handovers in this scenario have been extensively researched. Nevertheless, the handover mechanisms in LTE are inadequate in meeting the low latency requirements of 5G services. LTE-based handovers between small-cell APs require involvement of both CP and UP, resulting in longer latencies due to CP signaling between the small-cell AP and the CN. Furthermore, if dissimilar RATs are utilized between small-cell APs, additional signaling will be required.

The proposed HetNet system model accommodates the use of different RATs across small-cell APs. To support flexible deployment of small cells while reducing AP costs, only UP handovers are necessary between APs. The macrocell BS controls the handover between these APs, and each macrocell BS is equipped with an AMF that operates independent of RATs. Therefore, handovers between small-cell APs in 5G are RAT-independent. As handovers are handled by the macrocell BS, the procedure resembles that of a handover from a macrocell BS to a small-cell AP, as depicted in Figure 6.5.

When a handover is triggered, the UE sends a measurement report to the BS. The BS then makes a handover decision and communicates the handover request to the target and source APs via the Xn interface. Upon receiving the handover command from the BS, the UE synchronizes with the target AP and may obtain a new uplink allocation from the macrocell BS if required. As both small-cell APs remain under the coverage of the same macrocell, no CP handover is necessary. A UP handover between small-cell APs incurs

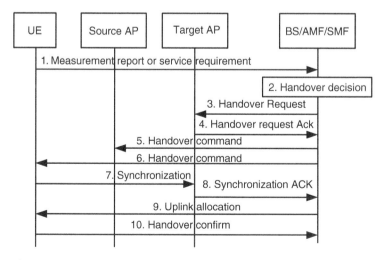

Figure 6.5 UP handover procedures between two small-cell APs.

lower signaling overhead and latency. All signaling messages related to the handover, as illustrated in Figure 6.5, occur within the RAN.

- Handover between two 5G macrocell BSs: 5G urban macrocells may use carrier frequencies around 2 GHz, 4 GHz, or up to 30 GHz, enabling wireless communication between 5G macro-BSs. These new BSs are designed for multi-RAT network deployments, with varying use cases that require high data rates and low latency. Figure 6.1 shows that handovers may occur between two 5G macro-BSs utilizing different RATs. Inter-RAT handovers in LTE involve significant signaling overhead due to the LTE CN. However, in 5G, the new CN architecture utilizing SDN and function separation simplifies inter-RAT handovers. As the AMF is RAT-independent, inter-RAT handovers follow the same procedure as those within the same RAT in 5G wireless networks. However, the handover process may differ based on whether the source 5G macro-BS has an unused authentication vector. Therefore, this subsection focuses on presenting the handover procedures between two 5G macro-BSs.

The handover procedures between two 5G macro-BSs are illustrated in Figure 6.6, assuming the source 5G macro-BS has at least one unused authentication vector. During these procedures, the UE may maintain a UP connection with a small-cell AP situated in the overlap area of the source and target macrocells. Because the source 5G macro-BS has at least one unused authentication vector, the handover process can be completed within the RAN. The source 5G macro-BS can utilize an unused authentication vector to establish mutual AKA between the UE and the target 5G macro-BS, as illustrated in Figure 6.6. In cases where the source 5G macro-BS has no unused authentication vector, the handover procedures between two 5G macro-BSs are presented in Figure 6.7. To complete the handover procedures, the 5G CN is required to generate the authentication vector.

From Figure 6.7, the communication overhead to generate the authentication vectors in CN can introduce significant latency. In order to prevent situations where a 5G macro-BS

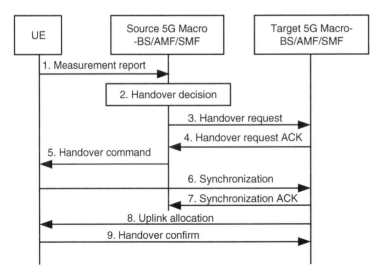

Figure 6.6 Handover procedures between two 5G macrocell BSs with unused authentication vector.

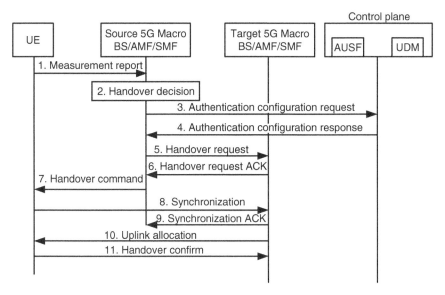

Figure 6.7 Handover procedures between two 5G macro-BSs without unused authentication vector.

does not have any unused authentication vectors during a handover, an authentication vector requirement can be established to ensure that each BS has at least one unused vector available. If a BS does not meet this requirement, it may request additional authentication vectors before proceeding with the handover in order to reduce latency.

6.4 A New Authentication Protocol for 5G Networks

To ensure secure mobility management in cellular networks, an AKA protocol is commonly used. The LTE AKA protocol achieves mutual authentication between a UE and the CN with key agreement [Ouaissa et al., 2018]. However, with the introduction of the next-generation CN in 5G networks, the AKA protocol also needs to be updated. Additionally, the LTE AKA protocol does not address the issue of preserving UE identity privacy, which is a significant concern in many use cases over 5G wireless networks. LTE AKA is known to have vulnerabilities such as MITM and DoS attacks, which are expected to be addressed in 5G AKA [Abdrabou et al., 2015]. The new features of 5G RAN and 5G CN will require different trust models, and hence new perspectives for authentication protocols. Although the 5GAKA draft was published in January 2018 as 3GPP TS 33.501 draft v0.7.0, which has not changed much from LTE AKA except for the entities involved, the authentication for 5G handover needs to be more efficient than LTE AKA to meet the advanced requirements of 5G services. To meet the service requirements of 5G, fast authentication protocols in addition to the full authentication protocol are necessary. Moreover, to enhance security against MITM and DoS attacks, the authentication protocol should also provide a confidentiality service. This section presents 5G AKA protocols for pre-authentication, full authentication, and fast authentication to provide the confidentiality service for authentication and to pre-authenticate the UE and the BS to the AUSF, without utilizing public key infrastructure.

6.4.1 Assumptions

In contrast to LTE, the new 5G authentication protocol uses dual identities for each UE. These include a real identity and a pseudo-identity. During the initial connection to the network, the UE sends its pseudo-identity over the air to the BS/AMF/SMF without encryption, as part of the pre-authentication process. Subsequently, the pseudo-identity is transmitted with encryption for enhanced security. To facilitate secure communication, the UE and the BS/AMF/SMF use a lightweight symmetric cryptographic function denoted by E. This function is shared between the UE and the network elements, enabling them to encrypt and decrypt messages exchanged during communication. Each UE and its home network (HN) AUSF share a key K', UE *PID*, and a hash function H_0. The BS/AMF/SMF and AUSF share a secret key K'' and a hash function H_1. The UE and its UDM share the UE PID, real ID, a long-term key K and two cryptographic algorithms H_2 and F, where H_2 are utilized to generate message authentication codes, and F is a key generation function.

For each AP, registration to the corresponding BS/AMF/SMF is needed for the first time to join the network. The registration can be done through the *Xn* interface, depending on the applications. The AP registration should achieve mutual authentication between the AP and the BS/AMF/SMF. A secret will be shared between the AP and the BS/AMF/SMF after the registration. This study does not delve into the specifics of AP registration. In scenarios where APs are located within the overlap area of BSs, they are required to register with the corresponding BSs. In cases where BSs have overlapping regions, a shared secret is used to secure the communication between them via the *Xn* interface.

6.4.2 Pre-Authentication

The LTE AKA protocol does not offer full confidentiality protection during the authentication process, exposing UE to potential eavesdropping, identity privacy breaches, MITM, and DoS attacks. As a result, a pre-authentication protocol is developed for each UE joining the network for the first time. Figure 6.8 illustrates the pre-authentication scheme, which comprises Steps 0-1 to 0-5. The dotted lines in the figure indicate unsecured links, while solid lines represent secured links.

The pre-authentication protocol is devised to accomplish two main objectives: first, to authenticate both the UE and the BS to the AUSF, and second, to establish a shared secret between the UE and the BS, which is used to provide end-to-end confidentiality protection throughout the entire authentication process. By fulfilling these requirements, the pre-authentication protocol enables secure and reliable UE access to the network. The following are the steps involved in the pre-authentication process.

- 0-1: The UE sends the pseudo-identity (PID), a time stamp (TS), and the identity of the home network ID_{HN} to the nearest BS/AMF/SMF.

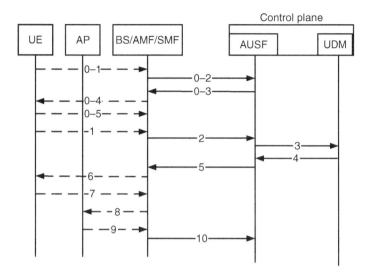

Figure 6.8 The proposed full authentication protocol.

- 0-2: After verifying the TS, the BS/AMF/SMF generates a message authentication code $H_1(K'', (PID \oplus ID_{HN}))$ using the key K'' and sends it, along with the PID and the Home Network Identity ID_{HN}, to the appropriate network entities for further processing.
- 0-3: The AUSF first verifies the message authentication code. If the message authentication code is verified, then the AUSF checks the PID on its list. If the PID and ID_{HN} are verified, then the AUSF generates a $RAND_0$ and computes message authentication codes $H_0(K', RAND_0)$ and $H_1(K'', (RAND_0 \oplus H_0(K', RAND_0)))$. The AUSF sends the $(RAND_0, H_0(K', RAND_0), H_1(K'', (RAND_0 \oplus H_0(K', RAND_0))))$ to the BS/AMF/SMF.
- 0-4: The BS/AMF/SMF verifies the message authentication code $H_1(K'', (RAND_0 \oplus H_0(K', RAND_0)))$. If the message authentication code is verified, the BS/AMF/SMF generates a $RAND_1$ and computes $[H_0(K', RAND_0)]^{RAND_1}$. The BS/AMF/SMF sends $(RAND_0, H_0(K', RAND_0), [H_0(K', RAND_0)]^{RAND_1})$ to the UE.
- 0-5: The UE verifies the $H_0(K', RAND_0)$, generates a $RAND_2$, computes $k_0 = [H_0(K', RAND_0)]^{RAND_1 RAND_2}$. The UE sends the $H_0(K', RAND_2)$ to the BS/AMF/SMF. And the BS/AMF/SMF verifies the hash value and computes $k'_0 = [H_0(K', RAND_0)]^{RAND_1 RAND_2}$, where $k_0 = k'_0$, which is the secret shared between the UE and the BS to provide confidentiality service for the full authentication protocol.

6.4.3 Full Authentication

Once the pre-authentication process is complete, every UE joining the network for the first time is required to undergo a full authentication procedure to establish a secure connection with the network. This is necessary to ensure the confidentiality and integrity of the communication and to prevent potential security breaches. The new full authentication protocol enables various security features, including mutual authentication between the UE and the BS, the UE and the UDM, and the UE and the AP. Additionally, it facilitates key generation and agreement for encryption and integrity protection. These measures are crucial for establishing a secure and reliable connection between the UE and the network. A batch authentication vector can be generated and stored in a BS/AMF/SMF to speed up the handover between BSs.

Figure 6.8 illustrates the updated full authentication protocol, which comprises Steps 1–10. These steps detail the process of establishing a secure connection between the UE and the network by enabling mutual authentication, key generation, and agreement for encryption and integrity protection. By utilizing the shared secret $k_0 = k_0'$ established during the pre-authentication process, the full authentication protocol can provide confidentiality services to ensure secure communication between the UE and the network. The follows describe the full authentication protocol in detail.

- 1: The UE generates a TS and encrypts the message of PID and TS with k_0', sends the $E_{k_0'}(PID, TS)$, the identity of home network ID_{HN}, the identity of AP ID_{AP} to the BS/AMF/SMF.
- 2: The BS/AMF/SMF decrypts the message with k_0, verifies TS, generates a message authentication code $H_1(K'', (PID \oplus ID_{HN} \oplus ID_{AP}))$, and sends PID, ID_{HN}, ID_{AP} with the message authentication code to the AUSF.
- 3: The AUSF verifies the message authentication code and sends PID to the UDM.
- 4: The UDM finds the corresponding real ID and long-term key K and generates a batch of m authentication vectors, each of them including a random number $RAND$, an authentication token $AUTH$, and an expected response ($XRES$). The UDM also generates a key K_{AUSF}. The UDM sends $(PID, AV[1, \ldots, m], K_{AUSF})$. In each authentication vector, the AUTN is represented as a message authentication code, denoted by $H_2(K, RAND)$. This code plays a crucial role in ensuring the integrity and authenticity of the authentication process by verifying the validity of the exchanged messages between the UE and the network.
- 5: The AUSF derives a key K_{AMF} from K_{AUSF} and sends K_{AMF} with $(PID, AV[1, \ldots, m])$ to the BS/AMF/SMF.
- 6: The BS/AMF/SMF generates two keys K_e and K_i for confidentiality and integrity protection for the link between the UE and the BS/AMF/SMF based on the key K_{AMF}, a $RAND'$ which is used to derive a session key K_{SMF}, a new pseudo ID (NPID). The BS/AMF/SMF randomly chooses an authentication vector from AV and sends the encrypted ($RAND$, $AUTH, RAND'$, and $NPID$) to the UE.
- 7: The UE verifies the $AUTH$. If $AUTH$ is verified, the UE computes the RES, K_e, and K_i for confidentiality and integrity protection for the link between the UE and the BS/AMF/SMF based on $RAND$, and other pair of K_e' and K_i' for confidentiality and integrity protection of the link between UE and the AP based on $RAND'$. The UE sends the RES to the BS/AMF/SMF.
- 8: The BS/AMF/SMF verifies the RES received from the UE. If the RES is verified, the BS/AMF/SMF sends encrypted K_{SMF} and $NPID$ with the secret shared between the AP and the BS/AMF/SMF.
- 9: The AP decrypts the message and generates the pair of K_e' and K_i' for confidentiality and integrity protection of the link between UE and the AP based on K_{SMF} and sends the encrypted NPID to the BS/AMF/SMF.
- 10: The BS/AMF/SMF verifies the encrypted $NPID$ and sends an authentication confirmation with $NPID$, PID, and ID_{AP} to the AUSF.

Figure 6.9 The new fast authentication protocol – APs handover.

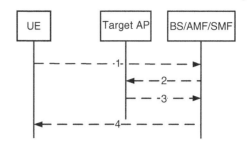

6.4.4 Fast Authentication

Fast authentication can be achieved in two scenarios, namely handover between APs and handover between BSs. In both scenarios, the primary goal is to establish a secure and reliable connection between the UE and the network as quickly as possible to ensure uninterrupted service and prevent potential security breaches.

6.4.4.1 Handover Between APs

As the small cells and APs join the network, they are registered with the BS/AMF/SMF, which establishes a shared secret between the APs and the BS/AMF/SMF to ensure secure communication. In fast authentication scenarios, the BS/AMF/SMF serves as a trusted authority to establish a new session for the UE during user plane handover. This approach ensures that the UE can transition seamlessly between different access points while maintaining the security and integrity of the connection.

Figure 6.9 illustrates the new fast authentication protocol, which does not involve the CN. As the UE maintains its connection with the BS/AMF/SMF, it can securely send the handover request to the BS/AMF/SMF. The fast authentication protocol consists of the following steps, which ensure the seamless transition of the UE between different access points while maintaining the security of the connection.

- 1: The UE sends the encrypted PID and a TS, and the identity of target AP ID_{TAP} to the connected BS/AMF/SMF for user plane handover.
- 2: The BS/AMF/SMF verifies the TS and sends the handover request to the target AP. The handover request includes an NPID and a session key K_{SMF} generated by the SMF. The handover request is encrypted by the secret shared between the BS/AMF/SMF and the target AP.
- 3: The target AP verifies the request, sends a confirmation to the BS/AMF/SMF, and generates a pair of confidentiality and integrity keys based on K_{SMF}.
- 4: The BS/AMF/SMF sends the confirmation to the UE with the NPID and the session key K_{SMF}. The confirmation message is encrypted with the secret shared between the UE and the BS/AMF/SMF. The UE generates a pair of confidentiality and integrity keys based on K_{SMF}.

When the NPID and the confidentiality and integrity keys are exchanged between the UE and the target AP, they can establish a secure communication channel and authenticate each other. This enables the UE to seamlessly transition to the target AP while maintaining the security of the connection. The BS/AMF/AMF will update the NPID to the CN after the UE connects with the target AP.

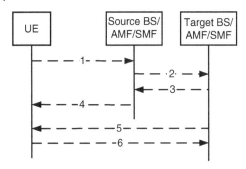

Figure 6.10 The new fast authentication protocol designed for handover between BSs.

6.4.4.2 Handover Between BSs

In addition to handovers between APs, this study also considers fast handovers between BSs. To achieve this, m batch authentication vectors are used to enable rapid authentication between BSs. During the BS handover, a small-cell AP maintains the UP connection with the UE, allowing for both CP and backup UP handovers to occur.

Figure 6.10 illustrates the new fast authentication protocol designed specifically for handover between BSs.

- 1: The UE sends the encrypted PID, the identity of the target BS ID_{TBS}, and a TS to the source BS/AMF/SMF for CP and backup UP handover.
- 2: The source BS/AMF/SMF verifies the TS, generates a session key K'_{SMF}, and sends it with the unused authentication vectors and the PID through a secure Xn interface to the target BS/AMF/SMF.
- 3: The target BS/AMF/SMF verifies the handover request and sends a confirmation to the source BS/AMF/SMF.
- 4: The source BS/AMF/SMF sends the encrypted session key K'_{SMF} to the UE.
- 5: The target BS/AMF/SMF generates a pair of keys for encryption and integrity protection based on the *RAND* in one of the unused *AV*. The target BS/AMF/SMF sends the encrypted *RAND* and *AUTH* from the unused authentication vector to the UE.
- 6: The UE verifies the message authentication code, computes the *RES*, a pair of keys for encryption and integrity protection based on *RAND*, and sends the encrypted *RES* to the target BS/AMF/SMF.

6.5 Security Analysis of the New 5G Authentication Protocols

This section provides a security analysis of the new authentication protocols with respect to authentication, confidentiality, and privacy properties.

Authentication: To prevent fake BS and MITM attacks during pre-authentication, it is essential to authenticate BS/AMF/SMF to AUSF using the shared secret as the key and the message authentication function. The UE also needs to authenticate itself to AUSF based on the shared hash function. Once AUSF has authenticated UE and BS/AMF/SMF in the pre-authentication, they can achieve secret sharing through Diffie–Hellman key exchange. The shared secret between UE and BS/AMF/SMF is then utilized to authenticate them both and provide confidentiality service for the full authentication. For full authentication, mutual authentication between UE and UDM is achieved through the long-term key shared between UE and UDM. The mutual authentication between UE and AP is achieved with the help of BS/AMF/SMF after the mutual authentication between UE and UDM. After the mutual authentication between UE and AP, the pair of confidentiality and integrity keys is shared between UE and AP. To expedite handover of BSs and APs without excessive signaling between the RAN and CN, multiple authentication vectors are created. The fast authentication process used during AP handover, which includes mutual authentication between UE and the target AP, is comparable to the full authentication process used for BS handover involving the BS, AMF, and SMF. The session key between UE and the AP is generated by BS/AMF/SMF. To expedite the authentication process during BS handover, mutual authentication between BSs is achieved using a shared secret. Additionally, unused authentication vectors can be leveraged by the target BS to establish new session keys, rather than being generated by the source BS. Overall, the new authentication protocols presented here enable mutual authentication between UE and BS, UE and CN, as well as UE and AP.

Confidentiality: Pre-authentication employs the Diffie–Hellman key exchange to facilitate secret sharing between UE and BS/AMF/SMF. This shared secret is leveraged to provide confidentiality during the full authentication protocol between UE and BS/AMF/SMF, effectively preventing eavesdropping and man-in-the-middle (MITM) attacks. After the mutual authentication between UE and UDM, the confidentiality of communications between UE and BS/AMF/SMF is provided by the confidentiality key generated based on the authentication vector. Confidentiality between UE and AP is established by generating a confidentiality key from the session key created by BS/AMF/SMF prior to mutual authentication. This confidentiality service effectively mitigates MITM and DoS attacks by ensuring that authentication parameters cannot be exploited.

Privacy: During pre-authentication, the PID of UE is transmitted in plaintext. However, after pre-authentication, both the PID and NPID are encrypted, which helps to safeguard the privacy of UE identities. The real identities of UE are only known by the UE and the CN. Each time UE associates with an AP, an NPID is assigned to the UE. Due to the proposed mechanism, the current AP is unable to access information about the previous or future AP associations of a UE. This ensures location privacy for the UE.

6.6 Performance Evaluations

This section presents an evaluation of the performance of the proposed authentication protocols in terms of communication overhead and computation overhead. To assess the communication overhead of the proposed authentication protocols, the total transmission message size for each protocol is calculated. Given that LTE AKA is employed during LTE handover scenarios, the performance of the fast authentication protocol is compared to that of LTE AKA. In terms of computation overhead, the same LTE AKA algorithms are utilized.

6.6.1 Communication Overhead

Table 6.1 presents the authentication parameters used for calculating communication overhead.

Table 6.2 illustrates a comparison of communication overhead between the proposed fast authentication and LTE authentication protocols. The proposed fast authentication protocols incur higher communication overhead between the UE and RAN, but do not require involvement of the CN. The signaling overhead between RAN and CN is significantly reduced. Overall, the proposed fast authentication protocols exhibit lower communication overhead. Additionally, due to the reduced distance between the UE and RAN, the fast authentication protocols are also more efficient in terms of communication costs and offer lower latency.

6.6.2 Computation Overhead

To compare computation overhead, the same algorithms used in LTE are employed. Cryptographic functions utilized in the authentication protocol, namely MAC, encryption/decryption, and key generation, is classified based on their computation requirements.

Table 6.1 Parameters of authentication protocols.

Symbol	Definition	Size (bits)
PID	Pseudo identity	128
NPID	New pseudo identity	128
ID_{HN}	Home network identity	128
ID_{BS}	Base station identity	64
ID_{AP}	Access point identity	64
K_{SMF}	Session management function key	128
K_{AMF}	Access and mobility management function key	256
RAND	Random number	128
TS	Time stamp	64
MAC	Message authentication code	64
RES XRES	Response expected response	64

Source: Adapted from Saxena et al. [2016].

Table 6.2 Communication overhead comparison.

Communication overhead (bits)	EPS AKA	Fast authentication for APs handover	Fast authentication for BSs handover
Between UE and RAN	627	640	960
CN side	1011	0	0

EPS, evolved packet system.

Table 6.3 Computation overhead comparison.

Computation overhead	EPS AKA	Fast authentication for APs handover	Fast authentication for BSs handover
UE side	B+C	2A+B+C	A+B+C
CN side	B+C	0	0

HMAC-SHA1 is used for MAC, 128-bit symmetric key encryption/decryption is used for encryption/decryption, and HMAC-SHA256 is used for key generation. Computation overhead for MAC, encryption/decryption, and key generation are represented by A, B, and C, respectively. Table 6.3 illustrates the comparison of computation overhead. The proposed fast authentication protocols entail no computation overhead on the CN or UE side. However, due to the additional confidentiality service provided in the full authentication protocol, the UE side incurs extra computation costs for encryption/decryption. Since the extra computation cost for encryption/decryption using symmetric algorithms is manageable, the proposed fast authentication protocols offer a viable solution for reducing computation and communication costs associated with handover processes. Furthermore, the absence of computation overhead on the CN side further contributes to lowering communication costs.

6.7 Conclusion

The aim of this chapter is to examine mobility management in 5G wireless networks, which demands secure and efficient management owing to the advanced communication and networking technologies employed in 5G, including HetNets and new radios. To overcome these challenges, a secure and efficient 5G mobility management scheme for SDN-based 5G wireless networks is proposed, leveraging the benefits of dual connectivity and separation of UP and CP. Various handover scenarios and schemes are analyzed under this proposed system model.

7

Open Issues and Future Research Directions for Security and Privacy in 5G Networks

Despite the progress made in 5G wireless security and privacy, there are still numerous challenges and issues that need to be addressed. This chapter aims to identify some of the open issues and outline potential research directions to enhance the security and privacy of 5G wireless networks.

7.1 New Trust Models

With the advanced features and services offered by 5G wireless networks, there is a growing need to apply these services to various vertical industries, such as smart grids, smart homes, vehicular networks, and m-health networks. In contrast to legacy cellular networks, trust models in 5G wireless networks may involve additional actors, including new entities with varying trust levels. Therefore, authentication mechanisms may need to be implemented between various actors with multiple trust levels [AB Ericsson, 2018, Wang et al., 2021a].

Research has been conducted on trust models for various use cases, as reported in Jellen et al. [2021]. For instance, in Eiza et al. [2016], the authors proposed a system model to ensure secure data transmission for vehicular communications over 5G wireless networks. The proposed system model comprises department of motor vehicles (DMV), trusted authority (TA), law enforcement agency (LEA), and vehicles. The trust model in 5G wireless networks is more complex than that in legacy cellular networks, particularly in scenarios involving multiple actors with varying trust levels. Given the massive number of devices in 5G networks, new trust models are required to improve security services for IoT use cases, including authentication. However, while there has been some research on trust models for specific use cases, such as vehicular communications, there remains a lack of a comprehensive trust model that incorporates all relevant actors, including devices and fusion centers, as observed in Labib et al. [2015]. For certain applications, networks may have various types of devices connected, some of which may be used solely for data collection while others may access the internet. As such, the trust requirements for each device may differ, necessitating corresponding trust models that cater to their unique security needs. These trust models may have varying security requirements to meet additional security demands. For instance, to meet high-security level demands, dual authentication mechanisms such as password and biometric authentication may be

5G Wireless Network Security and Privacy, First Edition. Dongfeng (Phoenix) Fang, Yi Qian, and Rose Qingyang Hu.
© 2024 John Wiley & Sons Ltd. Published 2024 by John Wiley & Sons Ltd.

necessary [Huawei, 2015]. In the case of m-health networks, a trust model was proposed in Zhang et al. [2017a], which addressed the privacy concerns and specified the relationships between the client, network management, and physician.

In conclusion, as 5G networks continue to enable new applications, novel trust models are required to cater to their security needs and demands.

7.2 New Security Attack Models

Based on recent research activities, the most commonly used attack model in physical layer security (PLS) involves a single eavesdropper equipped with a single antenna. However, the number of eavesdroppers in 5G wireless networks can be high, and they may utilize advanced technologies such as massive multiple-input and multiple-output (MIMO) [Chen et al., 2016a]. Furthermore, practical scenarios can involve different types of attacks [Li et al., 2020], and considering only one type of attack neglects the possibility of cooperation among jammers or eavesdroppers, making physical layer (PHY) security more complex. Although increasing the transmission power of the sender can mitigate jamming attacks, it can also increase the vulnerability to eavesdropping attacks. Therefore, there is a need to explore and develop more comprehensive attack models to enhance PLS in 5G wireless networks.

In addition, the adoption of software-defined network (SDN) and network functions virtualization (NFV) in the new service delivery model has resulted in more exposed vulnerable points [Dabbagh et al., 2015]. The separation of software from hardware means that the security of software is no longer solely reliant on the specific security attributes of the hardware platform [AB Ericsson, 2018]. As a result, the need for strong virtualization isolation has become increasingly crucial. To address this issue, network slicing was introduced in NGMN Alliance [2016] to provide isolated security. Furthermore, Luo et al. [2015a] proposed an effective vulnerability assessment mechanism for SDN-based mobile networks that uses an attack graph algorithm. They also presented a comprehensive security attack vector map of SDN.

New attack models in 5G wireless networks, enabled by emerging technologies and delivery models, pose significant challenges for security implementation compared to legacy cellular networks [Dai et al., 2020]. Despite this, there is currently limited research on these new attack models and potential solutions.

7.3 Privacy Protection

The increased usage of data in new 5G applications necessitates the transmission of vast amounts of sensitive data through the open network platforms, thereby raising concerns about potential privacy breaches [Huawei, 2015, Huang et al., 2020]. Therefore, the safeguarding of privacy is of utmost importance for enabling the implementation of various applications. Depending on the security requirements, privacy protection in different use cases can vary, encompassing concerns such as location privacy and identity privacy [Fang and Qian, 2020]. Various privacy concerns arise in 5G wireless networks due to the open network platforms, which can potentially result in the leakage of massive amounts of sensitive data. To address these concerns, different privacy protection measures are needed based on the specific use cases and security requirements. For instance, in healthcare applications, privacy protection is crucial to secure patients' data and identity. In this regard, [Zhang et al., 2017a] proposed a protocol that ensures secure data access and mutual authentication between patients and physicians. Similarly, for vehicular communications, privacy protection involves safeguarding the identity of a vehicle and the transmitted video contents. Moreover, to avoid location leakage in HetNets, [Farhang et al., 2015] proposed a differential private association algorithm. However, the sensing of the service type offered to a user may also pose a threat to user privacy, as discussed in Eiza et al. [2016] and Huawei [2015]. Therefore, differentiated quality of privacy protection should be provided based on the service type, while simultaneously ensuring user privacy.

The protection of privacy in 5G networks presents a significant challenge due to the massive volume of sensitive data transmitted and the open network platforms, which raise concerns about potential privacy leakage. Currently, encryption mechanisms are mainly used to implement privacy protection, but the encryption and decryption processes may affect other 5G service requirements such as latency and efficiency. Moreover, the use of powerful data analysis methods like machine learning adds complexity to privacy protection. However, data analysis can also be used intelligently to reduce the cost of encryption and protect sensitive dimensions before data transmission. For identity privacy, new management approaches should be considered, while multiple association mechanisms can enhance location privacy in different use cases. All these factors contribute to the challenge of providing adequate privacy protection in 5G wireless networks.

7.4 Unified Security Management

Although security features like access authentication and confidentiality protection are similar to those used in legacy cellular networks, a standard and basic set of security features is still needed for 5G wireless networks due to the various services, access technologies, and devices used [HUAWEI, 2016]. The unique security perspectives in 5G wireless networks, such as security management across heterogeneous access and for numerous devices, require a new security framework. For instance, the proposed security architecture in Chapters 5 and 6 introduces flexible authentication and handover between different access technologies, which necessitates defining security management for heterogeneous access to ensure flexibility for all access technologies. Additionally, security management for burst access behavior in many devices, such as IoT applications, needs to be studied to support efficient access authentication.

References

AB Ericsson. 5G security-enabling a trustworthy 5G system. *Ericsson white paper, Ericsson AB, Torshamnsgatan*, 21, 2018.

Dariush Abbasi-Moghadam, Vahid TabaTaba Vakili, and A. Falahati. Combination of turbo coding and cryptography in nongeo satellite communication systems. In *2008 International Symposium on Telecommunications*, pages 666–670. IEEE, 2008.

Emad Abd-Elrahman, Hatem Ibn-Khedher, and Hossam Afifi. D2D group communications security. In *2015 International Conference on Protocol Engineering (ICPE) and International Conference on New Technologies of Distributed Systems (NTDS)*, pages 1–6. IEEE, 2015.

M. A. Abdrabou, A. D. E. Elbayoumy, and E. A. El-Wanis. LTE authentication protocol (EPS-AKA) weaknesses solution. In *2015 IEEE Seventh International Conference on Intelligent Computing and Information Systems (ICICIS)*, pages 434–441, Dec 2015. doi: 10.1109/IntelCIS.2015.7397256.

Ibrahim Abualhaol and Steven Muegge. Securing D2D wireless links by continuous authenticity with legitimacy patterns. In *2016 49th Hawaii International Conference on System Sciences (HICSS)*, pages 5763–5771. IEEE, 2016.

Nadia Adem, Bechir Hamdaoui, and Attila Yavuz. Pseudorandom time-hopping anti-jamming technique for mobile cognitive users. In *2015 IEEE Globecom Workshops (GC Wkshps)*, pages 1–6. IEEE, 2015.

Mamta Agiwal, Abhishek Roy, and Navrati Saxena. Next generation 5G wireless networks: A comprehensive survey. *IEEE Communications Surveys & Tutorials*, 18(3):1617–1655, 2016.

Muhammad Naveed Aman, Kee Chaing Chua, and Biplab Sikdar. Mutual authentication in IoT systems using physical unclonable functions. *IEEE Internet of Things Journal*, 4(5):1327–1340, Oct 2017.

Jeffrey G. Andrews, Stefano Buzzi, Wan Choi, Stephen V. Hanly, Angel Lozano, Anthony C. K. Soong, and Jianzhong Charlie Zhang. What will 5G be? *IEEE Journal on Selected Areas in Communications*, 32(6):1065–1082, 2014.

Diego F. Aranha, Ricardo Dahab, Julio López, and Leonardo B. Oliveira. Efficient implementation of elliptic curve cryptography in wireless sensors. *Advances in Mathematics of Communications*, 4(2):169–187, May 2010.

5G Wireless Network Security and Privacy, First Edition. Dongfeng (Phoenix) Fang, Yi Qian, and Rose Qingyang Hu.
© 2024 John Wiley & Sons Ltd. Published 2024 by John Wiley & Sons Ltd.

Wade Baker, Mark Goudie, Alexander Hutton, C. David Hylender, Jelle Niemantsverdriet, Christopher Novak, David Ostertag, Christopher Porter, Mike Rosen, Bryan Sartin, et al. 2011 data breach investigations report. *Verizon RISK Team,* Available: www .verizonbusiness.com/resources/reports/rp_databreach-investigationsreport-2011_en_ xg.pdf, pages 1–72, 2011.

Neil Irwin Bernardo and Franz De Leon. On the trade-off between physical layer security and energy efficiency of massive MIMO with small cells. In *2016 International Conference on Advanced Technologies for Communications (ATC)*, pages 135–140. IEEE, 2016.

Y. Cao, N. Zhao, F. R. Yu, M. Jin, Y. Chen, J. Tang, and V. C. M. Leung. Optimization or alignment: Secure primary transmission assisted by secondary networks. *IEEE Journal on Selected Areas in Communications*, 36(4):905–917, 2018a. ISSN 0733-8716. doi: 10.1109/JSAC.2018.2824360.

Y. Cao, C. Q. Liu, and C. H. Chang. A low power diode-clamped inverter-based strong physical unclonable function for robust and lightweight authentication. *IEEE Transactions on Circuits and Systems I: Regular Papers*, 65(11):3864–3873, Nov 2018b.

Bin Chen, Chunsheng Zhu, Wei Li, Jibo Wei, Victor C. M. Leung, and Laurence T. Yang. Original symbol phase rotated secure transmission against powerful massive MIMO eavesdropper. *IEEE Access*, 4:3016–3025, 2016a.

Min Chen, Yongfeng Qian, Shiwen Mao, Wan Tang, and Ximin Yang. Software-defined mobile networks security. *Mobile Networks and Applications*, 21(5):729–743, 2016b.

Y. J. Chen, T. Hsu, and L. C. Wang. Improving handover performance in 5G mm-Wave HetNets. In *GLOBECOM 2017 - 2017 IEEE Global Communications Conference*, pages 1–6, Dec 2017. doi: 10.1109/GLOCOM.2017.8254624.

Mauro Conti, Nicola Dragoni, and Viktor Lesyk. A survey of man in the middle attacks. *IEEE Communications Surveys & Tutorials*, 18(3):2027–2051, 2016.

Mehiar Dabbagh, Bechir Hamdaoui, Mohsen Guizani, and Ammar Rayes. Software-defined networking security: pros and cons. *IEEE Communications Magazine*, 53(6):73–79, 2015.

Minghui Dai, Zhou Su, Ruidong Li, Yuntao Wang, Jianbing Ni, and Dongfeng Fang. An edge-driven security framework for intelligent Internet of Things. *IEEE Network*, 34(5):39–45, 2020.

Yansha Deng, Lifeng Wang, Kai-Kit Wong, Arumugam Nallanathan, Maged Elkashlan, and Sangarapillai Lambotharan. Safeguarding massive MIMO aided HetNets using physical layer security. In *2015 International Conference on Wireless Communications & Signal Processing (WCSP)*, pages 1–5. IEEE, 2015.

Xiaoyu Duan and Xianbin Wang. Fast authentication in 5G HetNet through SDN enabled weighted secure-context-information transfer. In *2016 IEEE International Conference on Communications (ICC)*, pages 1–6. IEEE, 2016.

Elena Dubrova, Mats Näslund, and Göran Selander. CRC-based message authentication for 5G mobile technology. In *2015 IEEE Trustcom/BigDataSE/ISPA*, volume 1, pages 1186–1191. IEEE, 2015.

Mahmoud Hashem Eiza, Qiang Ni, and Qi Shi. Secure and privacy-aware cloud-assisted video reporting service in 5G-enabled vehicular networks. *IEEE Transactions on Vehicular Technology*, 65(10):7868–7881, 2016.

ERICSSON. 5G security. *ERICSSON White Paper*, 2015.

Kai Fan, Yuanyuan Gong, Zhao Du, Hui Li, and Yintang Yang. RFID secure application revocation for IoT in 5G. In *2015 IEEE Trustcom/BigDataSE/ISPA*, volume 1, pages 175–181. IEEE, 2015.

Dongfeng Fang and Yi Qian. 5G wireless security and privacy: Architecture and flexible mechanisms. *IEEE Vehicular Technology Magazine*, 15(2):58–64, 2020.

Dongfeng Fang, Yi Qian, and Rose Qingyang Hu. Security for 5G mobile wireless networks. *IEEE Access*, 6:4850–4874, 2017a.

Dongfeng Fang, Yi Qian, and Rose Qingyang Hu. Interference management for physical layer security in heterogeneous networks. In *2017 IEEE 15th International Conference on Dependable Autonomic and Secure Computing*, pages 133–138. IEEE, 2017b.

Dongfeng Fang, Yi Qian, and Rose Qingyang Hu. Security requirements and standards for 4G and 5G wireless systems. *GetMobile: Mobile Computing and Communications*, 21(1):15–20, 2018.

Dongfeng Fang, Yi Qian, and Rose Qingyang Hu. Security analysis for interference management in heterogeneous networks. *Ad Hoc Networks*, 84:1–8, 2019a.

Dongfeng Fang, Shengjie Xu, and Hamid Sharif. Security analysis of wireless train control systems. In *2019 IEEE Globecom Workshops (GC Wkshps)*, pages 1–6. IEEE, 2019b.

Dongfeng Fang, Yi Qian, and Rose Qingyang Hu. A flexible and efficient authentication and secure data transmission scheme for IoT applications. *IEEE Internet of Things Journal*, 7(4):3474–3484, 2020.

Sadegh Farhang, Yezekael Hayel, and Quanyan Zhu. PHY-layer location privacy-preserving access point selection mechanism in next-generation wireless networks. In *2015 IEEE Conference on Communications and Network Security (CNS)*, pages 263–271. IEEE, 2015.

Y. Gao, H. Ma, D. Abbott, and S. F. Al-Sarawi. PUF sensor: Exploiting PUF unreliability for secure wireless sensing. *IEEE Transactions on Circuits and Systems I: Regular Papers*, 64(9):2532–2543, Sept 2017.

Samah A. M. Ghanem and Munnujahan Ara. Secure communications with D2D cooperation. In *2015 International Conference on Communications, Signal Processing, and their Applications (ICCSPA'15)*, pages 1–6. IEEE, 2015.

Global Mobile Suppliers Association. The road to 5G: Drivers, applications, requirements and technical development. *A GSA Executive Report from Ericsson, Huawei and Qualcomm*, 2015.

P. Gope, J. Lee, and T. Q. S. Quek. Lightweight and practical anonymous authentication protocol for RFID systems using physically unclonable functions. *IEEE Transactions on Information Forensics and Security*, 13(11):2831–2843, Nov 2018.

D. Adionel Guimaraes, G. H. Faria Floriano, and L. Silvestre Chaves. A tutorial on the CVX system for modeling and solving convex optimization problems. *IEEE Latin*

America Transactions, 13(5):1228–1257, May 2015. ISSN 1548-0992. doi: 10.1109/TLA.2015.7111976.

A. D. Harper and X. Ma. Mimo wireless secure communication using data-carrying artificial noise. *IEEE Transactions on Wireless Communications*, 15(12):8051–8062, Dec 2016. ISSN 1536-1276. doi: 10.1109/TWC.2016.2611581.

Debiao He, Sherali Zeadally, Neeraj Kumar, and Jong-Hyouk Lee. Anonymous authentication for wireless body area networks with provable security. *IEEE Systems Journal*, 11(4):2590–2601, Dec 2017.

Charles Herder, Meng-Day Yu, Farinaz Koushanfar, and Srinivas Devadas. Physical unclonable functions and applications: A tutorial. *Proceedings of the IEEE*, 102(8):1126–1141, Aug 2014.

Jiaqi Huang, Dongfeng Fang, Yi Qian, and Rose Qingyang Hu. Recent advances and challenges in security and privacy for V2X communications. *IEEE Open Journal of Vehicular Technology*, 1:244–266, 2020.

Huawei. 5G security: Forward thinking. 2015.

HUAWEI. 5G scenarios and security design, 2016.

Isabel Jellen, Joseph Callenes-Sloan, and Dongfeng Fang. Heterogeneous system model for security in e-health applications. In *2021 IEEE International Conference on Communications Workshops (ICC Workshops)*, pages 1–6. IEEE, 2021.

Ying Ju, Hui-Ming Wang, Tong-Xing Zheng, and Qinye Yin. Secure transmission with artificial noise in millimeter wave systems. In *2016 IEEE Wireless Communications and Networking Conference*, pages 1–6. IEEE, 2016.

Pardeep Kumar, An Braeken, Andrei Gurtov, Jari Iinatti, and Phuong Hoai Ha. Anonymous secure framework in connected smart home environments. *IEEE Transactions on Information Forensics and Security*, 12(4):968–979, Apr 2017.

Mina Labib, Sean Ha, Walid Saad, and Jeffrey H. Reed. A colonel blotto game for anti-jamming in the Internet of Things. In *2015 IEEE Global Communications Conference (GLOBECOM)*, pages 1–6. IEEE, 2015.

Mads Lauridsen, Lucas Chavarria Gimenez, Ignacio Rodriguez, Troels B. Sorensen, and Preben Mogensen. From LTE to 5G for connected mobility. *IEEE Communications Magazine*, 55(3):156–162, 2017.

Xiaohua Li and E. Paul Ratazzi. MIMO transmissions with information-theoretic secrecy for secret-key agreement in wireless networks. In *MILCOM 2005-2005 IEEE Military Communications Conference*, pages 1353–1359. IEEE, 2005.

Yao Li, Bipjeet Kaur, and Birger Andersen. Denial of service prevention for 5G. *Wireless Personal Communications*, 57(3):365–376, 2011.

Weiwei Li, Zhou Su, Kuan Zhang, Abderrahim Benslimane, and Dongfeng Fang. Defending malicious check-in using big data analysis of indoor positioning system: An access point selection approach. *IEEE Transactions on Network Science and Engineering*, 7(4):2642–2655, 2020.

J. Liu, Z. Zhang, X. Chen, and K. S. Kwak. Certificateless remote anonymous authentication schemes for wirelessbody area networks. *IEEE Transactions on Parallel and Distributed Systems*, 25(2):332–342, Feb 2014.

Madhusanka Liyanage, Ahmed Bux Abro, Mika Ylianttila, and Andrei Gurtov. Opportunities and challenges of software-defined mobile networks in network security. *IEEE Security & Privacy*, 14(4):34–44, 2016.

Shibo Luo, Jun Wu, Jianhua Li, Longhua Guo, and Bei Pei. Toward vulnerability assessment for 5G mobile communication networks. In *2015 IEEE International Conference on Smart City/SocialCom/SustainCom (SmartCity)*, pages 72–76. IEEE, 2015a.

Yijie Luo, Li Cui, Yang Yang, and Bin Gao. Power control and channel access for physical-layer security of D2D underlay communication. In *2015 International Conference on Wireless Communications & Signal Processing (WCSP)*, pages 1–5. IEEE, 2015b.

A. Maiti, I. Kim, and P. Schaumont. A robust physical unclonable function with enhanced challenge-response set. *IEEE Transactions on Information Forensics and Security*, 7(1):333–345, Feb 2012.

Akshatha M. Nayak, Pranav Jha, and Abhay Karandikar. A centralized SDN architecture for the 5G cellular network. *arXiv preprint arXiv:1801.03824*, 2018.

NGMN Alliance. 5G white paper. *Next generation mobile networks, white paper*, pages 1–125, 2015.

NGMN Alliance. 5G security recommendations package 2: Network slicing, 2016.

Nam-Phong Nguyen, Trung Q. Duong, Hien Quoc Ngo, Zoran Hadzi-Velkov, and Lei Shu. Secure 5G wireless communications: A joint relay selection and wireless power transfer approach. *IEEE Access*, 4:3349–3359, 2016.

NOKIA. Security challenges and opportunities for 5G mobile networks, 2017.

Frédérique Oggier and Babak Hassibi. The secrecy capacity of the MIMO wiretap channel. *IEEE Transactions on Information Theory*, 57(8):4961–4972, 2011.

Leonardo B. Oliveira, Diego F. Aranha, Conrado P. L. Gouvêa, Michael Scott, Danilo F. Câmara, Julio López, and Ricardo Dahab. TinyPBC: Pairings for authenticated identity-based non-interactive key distribution in sensor networks. *Computer Communications*, 34(3):485–493, Mar 2011.

Mariya Ouaissa, A. Rhattoy, and M. Lahmer. Analysis of authentication and key agreement (AKA) protocols in long-term evolution (LTE) access network. In *Advances in Electronics, Communication and Computing*, pages 1–9. Springer, 2018.

A. Özçelikkale and T. M. Duman. Cooperative precoding and artificial noise design for security over interference channels. *IEEE Signal Processing Letters*, 22(12):2234–2238, Dec 2015. ISSN 1070-9908. doi: 10.1109/LSP.2015.2472275.

Nisha Panwar, Shantanu Sharma, and Awadhesh Kumar Singh. A survey on 5G: The next generation of mobile communication. *Physical Communication*, 18:64–84, 2016.

J. Prados-Garzon, O. Adamuz-Hinojosa, P. Ameigeiras, J. J. Ramos-Munoz, P. Andres-Maldonado, and J. M. Lopez-Soler. Handover implementation in a 5G

SDN-based mobile network architecture. In *2016 IEEE 27th Annual International Symposium on Personal, Indoor, and Mobile Radio Communications (PIMRC)*, pages 1–6, Sept 2016. doi: 10.1109/PIMRC.2016.7794936.

Yi Qian, Feng Ye, and Hsiao-Hwa Chen. *Security in Wireless Communication Networks*. John Wiley & Sons, 2021.

Jian Qiao, Xuemin Sherman Shen, Jon W. Mark, Qinghua Shen, Yejun He, and Lei Lei. Enabling device-to-device communications in millimeter-wave 5G cellular networks. *IEEE Communications Magazine*, 53(1):209–215, 2015.

Zhijin Qin, Yuanwei Liu, Zhiguo Ding, Yue Gao, and Maged Elkashlan. Physical layer security for 5G non-orthogonal multiple access in large-scale networks. In *2016 IEEE International Conference on Communications (ICC)*, pages 1–6. IEEE, 2016.

Qualcomm. Leading the world to 5G, 2016.

M. E. S. Saeed, Q. Liu, G. Tian, B. Gao, and F. Li. Remote authentication schemes for wireless body area networks based on the internet of things. *IEEE Internet of Things Journal*, 5(6):4926–4944, Dec 2018.

Neetesh Saxena, Santiago Grijalva, and Narendra S. Chaudhari. Authentication protocol for an IoT-enabled LTE network. *ACM Transactions on Internet Technology (TOIT)*, 16(4):1–20, 2016.

Peter Schneider and Günther Horn. Towards 5G security. In *2015 IEEE Trustcom/BigDataSE/ISPA*, volume 1, pages 1165–1170. IEEE, 2015.

Ravindranath Sedidi and Abhinav Kumar. Key exchange protocols for secure device-to-device (D2D) communication in 5G. In *2016 Wireless Days (WD)*, pages 1–6. IEEE, 2016.

O. Semiari, W. Saad, M. Bennis, and B. Maham. Mobility management for heterogeneous networks: Leveraging millimeter wave for seamless handover. In *GLOBECOM 2017 - 2017 IEEE Global Communications Conference*, pages 1–6, Dec 2017. doi: 10.1109/GLOCOM.2017.8254681.

A. Sheikholeslami, D. Goeckel, H. Pishro-Nik, and D. Towsley. Physical layer security from inter-session interference in large wireless networks. In *2012 Proceedings IEEE INFOCOM*, pages 1179–1187, Mar 2012. doi: 10.1109/INFCOM.2012.6195477.

C. Shen and M. van der Schaar. A learning approach to frequent handover mitigations in 3GPP mobility protocols. In *2017 IEEE Wireless Communications and Networking Conference (WCNC)*, pages 1–6, Mar 2017. doi: 10.1109/WCNC.2017.7925950.

Yi-Sheng Shiu, Shih Yu Chang, Hsiao-Chun Wu, Scott C.-H. Huang, and Hsiao-Hwa Chen. Physical layer security in wireless networks: A tutorial. *IEEE Wireless Communications*, 18(2):66–74, 2011.

L. Sibomana, Hung Tran, and H. J. Zepernick. On physical layer security for cognitive radio networks with primary user interference. In *MILCOM 2015 - 2015 IEEE Military Communications Conference*, pages 281–286, Oct 2015. doi: 10.1109/MILCOM.2015.7357456.

SIMalliance. An analysis of the security needs of the 5G market, 2016.

Tianyi Song, Ruinian Li, Bo Mei, Jiguo Yu, Xiaoshuang Xing, and Xiuzhen Cheng. A privacy preserving communication protocol for IoT applications in smart homes. *IEEE Internet of Things Journal*, 4(6):1844–1852, Dec 2017.

Chris Sperandio and Paul G. Flikkema. Wireless physical-layer security via transmit precoding over dispersive channels: Optimum linear eavesdropping. In *MILCOM 2002. Proceedings*, volume 2, pages 1113–1117. IEEE, 2002.

William Stallings. *Cryptography and network security: Principles and practice*. Pearson, Upper Saddle River, NJ, 2017.

William Stallings, Lawrie Brown, Michael D. Bauer, and Michael Howard. *Computer security: Principles and practice*, volume 2. Pearson, Upper Saddle River, NJ, 2012.

Yipin Sun, Rongxing Lu, Xiaodong Lin, Xuemin Shen, and Jinshu Su. An efficient pseudonymous authentication scheme with strong privacy preservation for vehicular communications. *IEEE Transactions on Vehicular Technology*, 59(7):3589–3603, 2010.

S. Tang, L. Ma, and Y. Xu. A novel parameter estimation algorithm based on GHMM for vertical handover. In *2016 IEEE Global Communications Conference (GLOBECOM)*, pages 1–5, Dec 2016. doi: 10.1109/GLOCOM.2016.7841528.

Fengyu Tian, Peng Zhang, and Zheng Yan. A survey on C-RAN security. *IEEE Access*, 5:13372–13386, 2017.

Wade Trappe. The challenges facing physical layer security. *IEEE Communications Magazine*, 53(6):16–20, 2015.

Nils Ulltveit-Moe, Vladimir A. Oleshchuk, and Geir M. Køien. Location-aware mobile intrusion detection with enhanced privacy in a 5G context. *Wireless Personal Communications*, 57(3):317–338, 2011.

Chunzhi Wang and Yanmei Zhang. New authentication scheme for wireless body area networks using the bilinear pairing. *Journal of Medical Systems*, 39(11):136, Sept 2015.

H. M. Wang, C. Wang, T. X. Zheng, and T. Q. S. Quek. Impact of artificial noise on cellular networks: A stochastic geometry approach. *IEEE Transactions on Wireless Communications*, 15(11):7390–7404, Nov 2016a. ISSN 1536-1276. doi: 10.1109/TWC.2016.2601903.

Hui-Ming Wang, Tong-Xing Zheng, Jinhong Yuan, Don Towsley, and Moon Ho Lee. Physical layer security in heterogeneous cellular networks. *IEEE Transactions on Communications*, 64(3):1204–1219, 2016b.

W. Wang, K. C. Teh, and K. H. Li. Artificial noise aided physical layer security in multi-antenna small-cell networks. *IEEE Transactions on Information Forensics and Security*, 12(6):1470–1482, Jun 2017. ISSN 1556-6013. doi: 10.1109/TIFS.2017.2663336.

Yuntao Wang, Zhou Su, Qichao Xu, and Dongfeng Fang. Trusted and collaborative data sharing with quality awareness in autonomous driving. In *ICC 2021-IEEE International Conference on Communications*, pages 1–6. IEEE, 2021a.

Yuntao Wang, Zhou Su, Ning Zhang, and Dongfeng Fang. Disaster relief wireless networks: Challenges and solutions. *IEEE Wireless Communications*, 28(5):148–155, 2021b.

Dan Warren and Calum Dewar. Understanding 5G: Perspectives on future technological advancements in mobile. *GSMA Intelligence*, 2014.

Lili Wei, Rose Qingyang Hu, Yi Qian, and Geng Wu. Energy efficiency and spectrum efficiency of multihop device-to-device communications underlaying cellular networks. *IEEE Transactions on Vehicular Technology*, 65(1):367–380, 2016.

H. Xiong. Cost-effective scalable and anonymous certificateless remote authentication protocol. *IEEE Transactions on Information Forensics and Security*, 9(12):2327–2339, Dec 2014.

Min Xu, Xiaofeng Tao, Fan Yang, and Huici Wu. Enhancing secured coverage with CoMP transmission in heterogeneous cellular networks. *IEEE Communications Letters*, 20(11):2272–2275, 2016a.

Qian Xu, Pinyi Ren, Houbing Song, and Qinghe Du. Security enhancement for IoT communications exposed to eavesdroppers with uncertain locations. *IEEE Access*, 4:2840–2853, 2016b.

Shengjie Xu, Dongfeng Fang, and Hamid Sharif. Efficient network anomaly detection for edge gateway defense in 5G. In *2019 IEEE Globecom Workshops (GC Wkshps)*, pages 1–5. IEEE, 2019.

Qichao Xu, Zhou Su, Ruidong Li, Koichi Asatani, and Dongfeng Fang. Game theoretical secure bandwidth allocation in UAV-assisted heterogeneous networks. In *ICC 2021-IEEE International Conference on Communications*, pages 1–5. IEEE, 2021.

Qichao Xu, Zhou Su, Dongfeng Fang, and Yuan Wu. Hierarchical bandwidth allocation for social community-oriented multicast in space-air-ground integrated networks. *IEEE Transactions on Wireless Communications*, 22(3):1915–1930, 2022.

Ye Yan, Yi Qian, Hamid Sharif, and David Tipper. A survey on cyber security for smart grid communications. *IEEE Communications Surveys & Tutorials*, 14(4):998–1010, 2012.

L. Yan, X. Fang, and Y. Fang. A novel network architecture for C/U-plane staggered handover in 5G decoupled heterogeneous railway wireless systems. *IEEE Transactions on Intelligent Transportation Systems*, 18(12):3350–3362, Dec 2017a. ISSN 1524-9050. doi: 10.1109/TITS.2017.2685426.

W. Yan, F. Tehranipoor, and J. A. Chandy. PUF-based fuzzy authentication without error correcting codes. *IEEE Transactions on Computer-Aided Design of Integrated Circuits and Systems*, 36(9):1445–1457, Sept 2017b.

J. Yue, C. Ma, H. Yu, and W. Zhou. Secrecy-based access control for device-to-device communication underlaying cellular networks. *IEEE Communications Letters*, 17(11):2068–2071, Nov 2013. ISSN 1089-7798. doi: 10.1109/LCOMM.2013.092813.131367.

Alessio Zappone, Pin-Hsun Lin, and Eduard Jorswieck. Artificial-noise-assisted energy-efficient secure transmission in 5G with imperfect CSIT and antenna correlation. In *2016 IEEE 17th International Workshop on Signal Processing Advances in Wireless Communications (SPAWC)*, pages 1–5. IEEE, 2016.

Jianmin Zhang, Weiliang Xie, and Fengyi Yang. An architecture for 5G mobile network based on SDN and NFV. In *6th International Conference on Wireless, Mobile and Multi-Media (ICWMMN 2015)*, 2015.

H. Zhang, T. Wang, L. Song, and Z. Han. Interference improves PHY security for cognitive radio networks. *IEEE Transactions on Information Forensics and Security*, 11(3):609–620, Mar 2016. ISSN 1556-6013. doi: 10.1109/TIFS.2015.2500184.

Aiqing Zhang, Lei Wang, Xinrong Ye, and Xiaodong Lin. Light-weight and robust security-aware D2D-assist data transmission protocol for mobile-health systems. *IEEE Transactions on Information Forensics and Security*, 12(3):662–675, Mar 2017a.

Chensi Zhang, Jianhua Ge, Jing Li, Fengkui Gong, and Haiyang Ding. Complexity-aware relay selection for 5G large-scale secure two-way relay systems. *IEEE Transactions on Vehicular Technology*, 66(6):5461–5465, 2017b.

Zhenguo Zhao. An efficient anonymous authentication scheme for wireless body area networks using elliptic curve cryptosystem. *Journal of Medical Systems*, 38(2):13, Jan 2014.

N. Zhao, F. R. Yu, M. Li, Q. Yan, and V. C. M. Leung. Physical layer security issues in interference-alignment-based wireless networks. *IEEE Communications Magazine*, 54(8):162–168, Aug 2016. ISSN 0163-6804. doi: 10.1109/MCOM.2016.7537191.

N. Zhao, Y. Cao, F. R. Yu, Y. Chen, M. Jin, and V. C. M. Leung. Artificial noise assisted secure interference networks with wireless power transfer. *IEEE Transactions on Vehicular Technology*, 67(2):1087–1098, Feb 2018. ISSN 0018-9545. doi: 10.1109/TVT.2017.2700475.

Index

5G Wireless Network Security and Privacy, First Edition. Dongfeng (Phoenix) Fang, Yi Qian, and Rose Qingyang Hu.
© 2024 John Wiley & Sons Ltd. Published 2024 by John Wiley & Sons Ltd.